# Design and Analysis of Functionally Graded Adhesively Bonded Joints of FRP Composites

This book provides up-to-date information relevant to the analysis and design of adhesively bonded joints made up of fiber-reinforced polymer (FRP) composites using functionally graded adhesive (FGA). Damage behaviors in adhesively bonded joints of laminated FRP composites have been addressed, and joint configurations have been modeled using special finite elements (FEs) and multipoint constraint elements to simulate the contact behavior. Detailed 3D finite element analyses (FEAs) have been presented for different adhesively bonded joint structures along with guidelines for effective design philosophy of adhesively bonded joints in laminated FRP structures using FGA.

Features:

- Provides a thorough and systematic discussion on the functionally graded adhesive and its joints.
- Discusses analytical modeling and numerical analyses of the joints.
- Details 3D stress and failure delamination analysis for composite analyses of functionally graded out-of-plane joints under various combinations of loading.
- Illustrates FE modeling and simulation of interfacial failure and damage propagation in out-of-plane joints.
- Includes the effect of various gradation function profiles on damage growth driving forces characterized as Strain Energy Release Rate (SERR).

This book is aimed at researchers, professionals and graduate students in composites, infrastructure engineering, bonding technology and mechanical/aerospace engineering.

# Design and Analysis of Functionally Graded Adhesively Bonded Joints of FRP Composites

Sashi Kanta Panigrahi and Sunil V. Nimje

CRC Press
Taylor & Francis Group
Boca Raton London New York

CRC Press is an imprint of the
Taylor & Francis Group, an **informa** business

First edition published 2023
by CRC Press
6000 Broken Sound Parkway NW, Suite 300, Boca Raton, FL 33487-2742

and by CRC Press
4 Park Square, Milton Park, Abingdon, Oxon, OX14 4RN

*CRC Press is an imprint of Taylor & Francis Group, LLC*

© 2023 Sashi Kanta Panigrahi and Sunil V. Nimje

ISBN: 9781032061870 (hbk)
ISBN: 9781032061894 (pbk)
ISBN: 9781003201113 (ebk)

DOI: 10.1201/9781003201113

Typeset in Times
by codeMantra

# Contents

              Single Supported Tee Joint.................................................29

              3.1   Introduction .......................................................29
              3.2   Modeling and Analysis of Tee Joint.......................31
                    3.2.1   Failure Prediction.....................................36
                    3.2.2   Validation of Finite Element Model.........36
                    3.2.3   Convergence Study...................................37
              3.3   Design Optimization of Tee Joint...........................37
                    3.3.1   Lamination Schemes and Material Anisotropy .........37
                    3.3.2   Stress Analysis of Tee Joint with Functionally
                            Graded Adhesive .....................................41
              3.4   Summary ............................................................44

**Chapter 4**   Study of Damage Growth in Functionally Graded Adhesively
              Bonded Double Supported Tee Joint.................................47

              4.1   Introduction .......................................................47
              4.2   Numerical Simulation of Double Supported Tee Joint
                    Using Finite Element Analysis ................................50
              4.3   Interfacial Failure Propagation Analysis in Tee Joint ............53
              4.4   Effect of Material Gradation of Adhesive on Damage Growth .........59
              4.5   Summary ............................................................61

**Chapter 5**   Functionally Graded Adhesively Bonded Joint Assembly under
              Varied Loading.................................................................63

              5.1   Introduction .......................................................63
              5.2   Analysis and Evaluation of Different Loadings on the
                    Failures of Tee Joint .............................................65
              5.3   Cohesive and Interfacial Failure Onset in Tee Joint Structure...........69
              5.4   Influence of Graded Bond Line on-resistance of
                    Joint Assembly....................................................69
                    5.4.1   Tee Joint under Tensile Loading ...............70
                    5.4.2   Tee Joint under Compressive Loading ......73
                    5.4.3   Tee Joint under Bending Loading .............75
                            5.4.3.1   Vertical Bond line.....................75
                            5.4.3.2   Horizontal Bond line ................77
              5.5   Summary ............................................................78

**Chapter 6**   Effects of Functionally Graded Adhesive on Failures of Tubular
              Lap Joint of Laminated FRP Composites .........................81

              6.1   Introduction .......................................................81
              6.2   Structural Behavior of Tubular Joint under Pressure
                    and Axial Loads ...................................................83

# Authors

**Sashi Kanta Panigrahi** (PhD, IIT Kharagpur) is working as a Professor in the Department of Mechanical Engineering of Defence Institute of Advanced Technology (Deemed to be University), Pune and, currently holding the position of Dean (Sponsored Research) and Director of International Co-operation Cell as an additional responsibility. Prior to this, he worked as Dean (Students Affairs) and Head of the Department for the Department twice along with many other important responsibilities. He has worked as an International Visiting Academic with the University of New South Wales at the Australian Defence Force Academy (UNSW@ADFA). He has more than 30 years of wide and intensive teaching, research, training and administrative experience. His research works primarily in the areas of analysis and design of composite materials, characterization of FRP composite materials, FEA of FRP composite materials and composite structures, natural fiber-reinforced composite (NFRC) materials, fracture mechanics principle applicable to modeling and simulation of damages in orthotropic and isotropic materials and material characterization/stress analysis/solid mechanics/machine design. He has been working on the development of advanced FE methods and nonlinear FEAs and modeling of engineering structures with functionally graded/monolithic adhesively bonded joints. He has published more than 210 research articles in peer-reviewed scholarly research papers and international journals/conferences including six books, one monograph and many conference proceedings including a series of lecture materials. He is a Fellow of Indian Society of Mechanical Engineers (ISME); Fellow of Institution of Engineers (IE), India; Member of Indian Association for Mechanism and Machines (AMM); and Member of Indian Academy of Computational Mechanics (IndACM) and many other professional bodies. He is the recipient of many awards such as Distinguished Scientist in Composite Structures Award by Venus International Research Foundation, Chennai, 2018; Innovative Technological Research & Dedicated Professor Award by JETR-JETMS Kuala Lumpur, Malaysia, 2017; and Bharat Jyoti Award, 2012. He has been engaged as a frequent reviewer for many leading peer-reviewed international journals of the highest standards. He has also served as a technical committee member or advisory board member for several national/international conferences and an active editorial board member of a couple of international journals.

**Sunil V. Nimje** is an Assistant Professor in the Department of Mechanical Engineering of Defence Institute of Advanced Technology (DIAT), Pune. He earned his MTech in Mechanical System Design from IIT, Kharagpur and PhD from DIAT, Pune. He has more than 15 years of teaching, research and industry experience. His research areas are in the field of design and analysis of composite structures, FE modeling and simulation of damages in orthotropic and isotropic materials, composite patch repairs and monolithic/functionally graded adhesively bonded joints. He has contributed to the field of damage and failure analysis of composite and functionally graded structures by authoring more than 50 research articles in peer-reviewed international journals/conferences including four book chapters. He has been engaged as a reviewer for peer-reviewed international journals. He is a life member of the Indian Society for Theoretical and Applied Mechanics (ISTAM).

His industry experience includes design and development of Dual Purpose Improved Conventional Munition (DPICM) Bomblet Fuze (Impact/Self Destruction) for nonspinning projectiles at DRDO. He has also been associated with Design and Development of Safety Arming Mechanisms for soft target munitions and anti-tank mines. He also worked as an Engineer at GE Aviation and was involved in fatigue life estimation of various rotating parts of aircraft engine.

# Preface

In recent years, the adhesively bonded joint structures made from laminated FRP composites are used extensively and are more popular in many applications, including aerospace, ground transport, civil infrastructure and maritime. The advantage of adhesively bonded joints over bolted or riveted joints is that the use of fastener holes in mechanical joints inherently results in micro and local damages to composite laminate during their fabrication. The adhesively bonded joints exhibit a common problem, called edge effects, arising from peel and shear stress concentrations occurring around the free edges of plate/tube-adhesive interfaces and affecting the overall joint strength considerably. Some geometry-specific measures were considered for adjusting stiffness of joint members around these free edges to improve joint strength. Functionally graded materials (FGMs) appear in nature with the role of reducing stress concentrations along the bi-material interfaces. In this book, the innovative concept of FGM is adopted by using FGA in lieu of conventional monolithic adhesive material for smoothing/relieving stress distributions over the entire bond line. Hence, design of efficient adhesively bonded joints using FGA is a critical issue for engineers/designers/researchers/practitioners.

Generally speaking, adhesively bonded joint structures having laminated FRP composite adherends joined using functionally graded adhesive pose a severe threat in terms of different types of damages. Usually, damages in adhesively bonded laminated FRP composites may manifest themselves in the forms of cohesive failure, interfacial failure, inter-/intralaminar delamination, debonding of fiber–matrix interface, matrix cracking and fiber breakage, etc. All or some of the above damage modes may simultaneously be present in any structurally bonded joint applications. Damages due to the cohesive failure, the interfacial failure and the delamination are of prime concern because they may reduce significantly the strength and stiffness of the structure leading to the loss of structural integrity and stability and even it may lead to final catastrophic failure. In this book, these damage behaviors in adhesively bonded joints of laminated FRP composites have been addressed in detail using the numerical technique based on three-dimensional (3D) FEA. The joint configurations have been modeled using special FEs, and multipoint constraint elements have been used for the interfacial damaged surfaces. The onset of damage and its propagation/growth in the joint can be assessed by the values of strain energy release rates (SERRs). The three components of SERR, viz. $G_I$, $G_{II}$ and $G_{III}$ (SERR in modes I, II and III), have been used as characterizing parameters for assessing the damage growth behavior. The modified crack closure integral techniques based on the concepts of linear elastic fracture mechanics have been used to evaluate the individual modes of strain energy release rates during the interfacial adhesion failure propagation. 3D analyses are essential for analyzing various types of damages. In this book, the detailed 3D FEAs have been presented for different adhesively bonded joints in FRP-laminated composites.

The objective of this book is to formulate and facilitate guidelines, which will enable to establish a better and more effective design philosophy of adhesively bonded joints in laminated FRP structures using functionally graded adhesive. The objectives of the present work have been set as follows.

Chapter 1 deals with an introduction and overview of laminated FRP composites, functionally graded adhesives, configurations of adhesively bonded joints with laminated FRP composites, and damages commonly observed in bonded joints. Challenges and prospects pertaining to manufacturing of functionally graded bonded joints are also addressed. Chapter 2 highlights the methodology for modeling and simulation of adhesion failure and the procedure for evaluation of SERR for characterizing the damage initiation and propagation in the joint structure. Chapter 3 is devoted to 3D stress analyses of functionally graded adhesively bonded joint. The appropriate material with a specific lamination scheme and material gradation function profile of adhesive based on stress and failure analyses are established for the joint designer in this chapter. In Chapter 4, the effect of pre-embedded adhesion failures in functionally graded adhesively bonded joints on the joint strengths, SERR variations and adhesion failure propagations are studied. The novelty of functionally graded adhesive for improved damage growth resistance of tee joint structure is exhibited. Chapter 5 demonstrates the behavior of tee joint bonded assembly under varied loading, using nonlinear FE method and highlights the idea of improving the resistance of this assembly by using a functionally graded adhesive. SERR-based damage analyses of functionally graded adhesively bonded tubular lap joint structure under varied loadings using 3D geometrically nonlinear FEAs have been studied in Chapter 6. Design and analysis aspects of graded tubular joints are the main focus of Chapter 7. The prediction of the onset of adhesion failure and their propagations in an adhesively bonded laminated FRP composite socket joint have been presented. The influence of graded adhesive in arresting the damage growth is highlighted in this chapter.

This book will serve the needs at graduate and postgraduate levels as a reference book as well as the needs of practicing engineers and scientists working in academics, research and development establishments, defense sectors, aerospace industries and marine applications. It is hoped that this book will help provide impetus for teaching advanced courses on damages in graded bonded joints at universities as well as support short courses for professional development of engineers in industry. The wealth of material covered can also help new researchers in advancing the field further.

# 1 Structural Bonded Joint with Functionally Graded Adhesive

## An Introduction and Overview

## 1.1 BACKGROUND

The demand for composite joints is increasing day by day in many applications such as aerospace, ship designs, wind turbines, transport, chemical and automotive. The current emphasis of the composite design is on increased performance with reduced material and manufacturing costs. Composites are commonly used because of their high strength and stiffness, low mass, excellent durability and ability to form complex shapes. The efficient transfer of load through the assembly of complex composite structures requires an efficient joining method. These joining methods can be achieved in different ways by bolting, riveting, brazing, soldering, welding of metallic (isotropic) components or by adhesive bonding of both isotropic and orthotropic materials. One of the significant advantages of adhesive bonding is that it enables dissimilar materials to be joined, even when one of the components is non-metallic. In addition, it allows for uniform diffusion of the load into the structure, thus reducing the localized stresses encountered compared with other joining methods, mainly mechanical fastening. Therefore, a significant application of adhesive bonding is found in the joining of fiber-reinforced polymeric (FRP) composite materials widely used in composite structures. No matter what forms of connections are used in any system, the joints are potentially considered the weakest points. By using FRP composite materials, these weakest points increase, which may lead to the loss of structural integrity of the structure. Thus, adhesively bonded structural joints of FRP composite materials must be designed appropriately to meet the specific design requirements.

All the adhesively bonded joints present stress peaks or stress singularities at the edges of the bond line due to elastic mismatch or peel stresses, both in-plane and out-of-plane joints. Several approaches were proposed in the literature to mitigate and decrease the degree of the singularity of these peaks, such as relief grooves [1–3], scarf joints or rounding edges [4–6]. Material grading occurs naturally at material interfaces to reduce stress concentrations [7–9]. Biological interfaces such as tendon-to-bone joints have been found to have graded material properties to distribute stress more evenly across the joint [10]. Generally speaking, material grading has been explored by

DOI: 10.1201/9781003201113-1

continuously varying the elastic modulus of the adhesive by introducing functionally graded adhesive (FGA) for improved structural performance of the out-of-plane joints.

## 1.2  FIBER-REINFORCED POLYMERIC COMPOSITES: CONSTITUENTS AND CHARACTERISTICS

Composite materials are composed of two or more materials on a macro-scale resulting in a macroscopically homogeneous medium. Fiber-reinforced composite materials consist of fibers of high strength and modulus embedded in or bonded to a matrix with the distinct interface between them. Both fibers and matrix retain their physical and chemical identities in this form. Thus, FRP composites exhibit the best qualities compared with their constituents and often possess the better qualities that neither of their constituents possesses. In general, fibers are the principal load-carrying members. At the same time, the matrix keeps them at the desired location, and orientation acts as a load transfer medium between them and protects them from environmental damages. Even though the fibers provide reinforcement to the matrix, the latter also serves as a member of metal functions [11–14] in FRP composite materials.

The FRP composite materials are used in a large scale due to their improved properties such as specific strength, specific weight, stiffness, corrosion resistance, wear resistance, fatigue life, temperature-dependent behavior, thermal insulation, thermal conductivity and acoustical insulation. Naturally, not all of these properties are improved simultaneously nor is there usually any requirement to do so. Due to the above-desired characteristics, FRP composite materials find wide-scale applications in the industries such as aircraft, automotive, marine, sporting goods, biomedical sciences, electronics and defense. Unlike traditional monolithic materials, FRP composites can have their strengths oriented to meet specific design requirements of applications. Actual composite structures consist of multilayered laminae having different ply orientations of continuous fiber as shown in Figure 1.1.

A wide variety of fibers and matrix materials are now available for use, and the principal fibers in commercial use are various types of glass, carbon, graphite, aramid, etc. Other fibers such as boron, silicon carbide and aluminum oxide are used in limited applications. All these fibers are incorporated into the appropriate matrix phase in continuous or discontinuous (chopped) lengths. The matrix material may be a polymer, a metal or a ceramic. Specific fillers, additives and core materials are sometimes added to enhance and modify the final product.

The fibrous composite material is one of the most important ones from the application point of view. The filamentary type of composite material is that material system consisting of selected fiber-macro constituents. Such material systems have desirable properties, which are discussed above. The classification of such fibrous composite systems is shown in Figure 1.2.

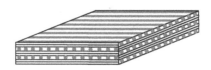

**FIGURE 1.1**  An example of a continuous fiber-laminated fiber-reinforced polymeric composite.

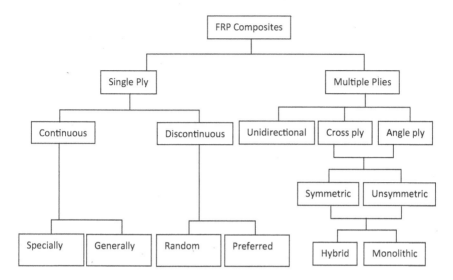

**FIGURE 1.2** Classification of fiber-reinforced polymeric composites.

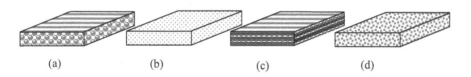

(a)       (b)       (c)       (d)

**FIGURE 1.3** Various types of polymeric composites: (a) fibrous, (b) particulate, (c) laminated and (d) flake.

Schematics of various types of composite laminates used for structural applications are shown in Figure 1.3. The essential constituents of the composite material system are the reinforcements, the matrix (usually epoxy or polyester) and the coupling agents (coatings/fillers).

Depending on the reinforcement type, they are classified as follows:

- **Fibrous:** composed of continuous or chopped fibers
- **Particulate:** composed of particles
- **Laminated:** composed of layers or laminae with desired fiber orientation
- **Flake:** composed of flat flake reinforcement

The selection of a specific composite material system must be carefully chosen to enhance the structural integrity and efficiency of any composite structure. The composite must be resistant to debonding or delamination damages at the fiber/matrix interface, and it must be resistant to fiber breakage and matrix cracking. However, in applications where it is desired to dissipate energy during the failure process (such as crashworthy or impact-resistant structures), progressive fiber failure and fiber–matrix debonding are positive features because they dissipate energy. Thus, a significant challenge for the mechanism and material community is understanding the factors influencing damage development and its propagation in structural applications. The different types of damages, propagation of damages and the mechanics of initiation of damages are discussed vividly in the subsequent sections.

Damages in adhesively bonded laminated FRP composites may manifest themselves in the forms of cohesive failure, interfacial failure, interlaminar/intralaminar delamination, debonding of fiber–matrix interface and fiber breakage. Although all or some of the above damage modes may simultaneously be present in any structural bonded joint applications, damages due to the cohesive failure, interfacial failure and delamination are of prime concern because they may significantly reduce the strength and stiffness of the structure. So, this has been a significant concern on engineering applications of the adhesively bonded joints because of structural integrity and stability problems, load-carrying capacity, stiffness reduction, exposure of the interior to an adverse environment and final catastrophic failure the structure. Delamination failure mode in the laminated FRP composite plates/tubes is especially insidious. FRP composites exhibit relatively poor resistance to delamination damages. These damages may arise from microcracks and cavities or voids formed during manufacturing stages, service or maintenance induced damages or from low-velocity impact damage [15–18]. Their susceptibility to out-of-plane loadings, such as transverse loading, is because matrix constituents mainly control properties in a transverse direction. Strain energy release rate (SERR) calculations along the straight or curved delamination fronts have been evaluated by many researchers [19–23] for a wide range of problems on FRP composites to assess the delamination or debonding growth.

On the whole, most of the analyses performed to date have been restricted to simple models for studying the delamination damages. However, the general problem of modeling the interfacial/delamination of any shape when existing in laminated FRP composites of adhesively bonded joints in composite structures can be tackled effectively by using a rigorous three-dimensional analysis. The reason why a full three-dimensional FEA is used, despite its high computational cost in terms of space and memory requirements, is to keep the study as general and versatile as possible. This would require very few embedded assumptions compared with 2D FEA and closed-form analytical formulations thereby the modeling and simulations would be very close to the real-life situation.

## 1.3   JOINING TECHNIQUES OF FIBER-REINFORCED POLYMERIC COMPOSITE LAMINATES

Adhesive bonding represents one of the essential enabling technologies for developing innovative design concepts and structural configurations and exploiting new materials. The evolution of the adhesive bonding method and its current knowledge was made possible by the explosive growth in adhesive applications in a great variety of industries over the past few decades. Although it is easy for everyone to identify examples of adhesive bonding in the world around us, analysis and design of structural bonded joints represent one of the challenging jobs in terms of analysis, design and structural integrity assessment. Compared with other mechanically fastened joints, adhesively bonded joints can offer substantially improved performance and economic advantages, which are listed below:

- Joining of dissimilar materials
- Continuous bond
- Stronger and stiffer joints
- More uniform stress distribution (shown in Figure 1.4) in the joint cross-section
- Low local stress concentrations
- Bonding of porous adherends possible
- No finishing costs
- Improved fatigue resistance
- Vibration damping and no noise
- Reduced weight and part count
- Bonding of large areas (both planar and non-planar)
- Accommodation of small areas for bonding
- Availability of fast and slow curing systems
- Easy to combine with other fastening methods such as pin/rivet/bolted joints
- Easily automated/mechanized

Analysis and design of adhesively bonded joints is a multidisciplinary task, and it can involve concepts from surface and polymer chemistry, stress analysis, manufacturing technology and fracture mechanics. The following essential aspects need to be considered for designing the adhesively bonded joints when used in structural applications:

- Selection of suitable adhesive and its characteristics
- Appropriate surface preparation of plates/tubes of bonded joint
- Development of design based on stress and failure analyses

## 1.4 ADHESIVELY BONDED JOINT CLASSIFICATIONS AND INDUSTRIAL APPLICATIONS

Broadly there are two types of joint configurations used in many applications, namely in-plane and out-of-plane joints. Most of the studies on adhesively bonded joints reported so far have focused on in-plane joints such as single lap, double lap, lap shear and butt joint. However, out-of-plane joints have been used extensively in many structural applications. A tee joint made up of plates/tubes is considered an

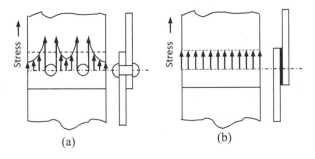

(a)                                    (b)

**FIGURE 1.4** Comparison of stress distribution in (a) mechanically fastened joint and (b) adhesively bonded joint.

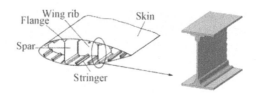

**FIGURE 1.5** Tee section stiffening skin element in aerofoil structure [26].

out-of-plane joint formed when a right-angled plate is adhesively bonded to a base plate. Multiple watertight bulkheads are used in a ship/ structure to divide the hull into many compartments by using tee joints. These sections are primary structures in maintaining the ship stiffness under various loading conditions. Tee joint is used to bond the bulkheads and hull to transfer loads and maintain watertight integrity. Then the reliability of the ship heavily depends on this connection between the substructures [24,25]. Similarly, T-stiffeners are used extensively in aircraft wings to provide strength during wing loading (Figure 1.5). The tee joints are needed to give rigidity to thin and relatively flexible composite plates [26].

Another category of out-of-plane joints is adhesively bonded tubular/pipe structures. Application of FRP composites for piping systems was developed in response to significant corrosion and wear problems associated with metallic pipes in the chemical process, pulp and paper, offshore oil and gas industries, etc. Pipes made from FRP composites have been widely used in wastewater treatment, power, petroleum, aerospace and automotive industries, etc. The complex layout of industrial piping systems, along with limitations associated with composite pipe manufacturing, demands repeatable and durable joining mechanisms. Composite tubes have been used in forming truss structures of space launch vehicles to reduce their weight. Mechanical joining methods of composite tubes such as trimming, bolting and fastening enhance stress raisers in the joint structure.

Currently, there are a large variety of out-of-plane bonded joint configurations used in many structural applications. Few adhesively bonded joint configurations used in practice are shown in Figure 1.6.

## 1.5 CHARACTERIZATION OF BONDED JOINT CONSTITUENTS

Any rational design of structural bonded joints must be based on adequate knowledge of the stresses and strength in the joint. To determine the stresses and further to predict the performance (strength, stiffness, service life, etc.) of the bonded joint, it is inevitable to know the material characterization of the constituents. In general, the bonded joint is composed of two distinct constituents: (i) the adhesive and (ii) plates/tubes as the adherend. It is necessary to discuss the material characterization of bonded joint constituents.

### 1.5.1 ADHESIVE

According to the need, adhesive materials are considered to be isotropic ones. Young's modulus and Poisson's ratio are the two-input data for linear stress analyses. In contrast, for non-linear analyses, stress–strain curves may be considered, and

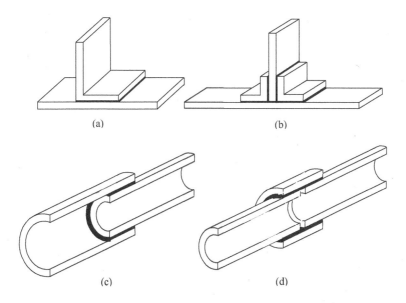

(a)                              (b)

(c)                              (d)

**FIGURE 1.6** Different types of adhesively bonded out-of-plane joints of laminated fiber-reinforced polymeric composites: (a) single supported tee joint, (b) double supported tee joint, (c) tubular lap joint and (d) tubular socket joint.

material yielding and hardening rules may also be needed. Fracture toughness and fatigue properties are required to characterize the adhesive's failure strength and service life, respectively.

Researchers have developed various test methods to characterize the adhesive material when sandwiched between the laminated FRP composite plates/tubes. Accurate test methods are published in ASTM (American Society for Testing and Materials) standards, BS (British Standard), ISO (International Standards Organization) standards and the EU (European) standard. These tests evaluate the material properties of adhesive such as strength, modulus and fracture toughness and the bonding techniques, effectiveness of surface preparations and curing cycle. All tests are classified into four groups, that is, shearing, tension, peeling and fracture toughness. The details of these tests are described in the literature [27,28].

Adhesive materials can be classified into (i) brittle adhesive and (ii) ductile adhesive. For a brittle adhesive, a proportional linear relationship exists between the stress and the strain, whereas for a ductile adhesive, a non-linear stress–strain relationship is generally observed, as shown in Figure 1.7. Young's modulus and Poisson's ratio are the two properties required for linear analyses of bonded joints for brittle adhesives.

### 1.5.2   PLATES/TUBES OF BONDED JOINT

One of the significant advantages of adhesive bonding is that it enables orthotropic materials to be joined, even though it can be used for metallic plates/tubes or a combination of both. In many applications, plates/tubes of bonded joints are made of laminated FRP composites. The analysis techniques are essentially the same as when

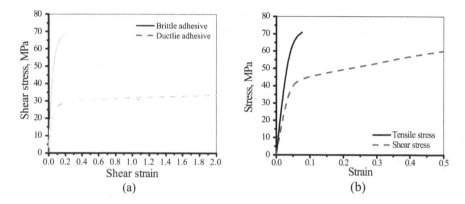

**FIGURE 1.7** Comparison of characteristics of adhesives: (a) shear stress–strain for brittle and ductile adhesives and (b) true shear and tensile stress–strain curves of adhesive [29].

isotropic plates/tubes are used, although due attention must be paid to the low longitudinal shear stiffness of unidirectional composites. With unidirectional composites, the shear modulus is of the order of 25%–30% of Young's modulus. It may be as low as 2%, so the shearing phenomenon in plate/tube becomes extremely important. Using lamination techniques in which fibers are placed at a different orientation to the plate/tube axis leads to reduced longitudinal and increased shear moduli. However, the transverse modulus remains low. In addition, the transverse strength of FRP composite plate/tube is low, usually being the same order or less than that of the matrix. Thus, if the joint experiences transverse (peel) loading, there is a strong likelihood that the composite will fail in transverse tension before the adhesive fails. Therefore, adhesive peel stresses should be minimized when composite plates/tubes are used, lest this leads to plate/tube failure.

Compared with isotropic materials, the analysis of joints between FRP composites is complicated by the anisotropy and heterogeneity of the plates/tubes. A rigorous analysis may also be carried out to include the effects of residual thermal strains arising from curing and thermal mismatch when bonding to metals. The additional variables such as lay-up scheme, ply orientation and stacking sequence will significantly affect the performance of the joint made of laminated FRP composite plates and tubes. These parameters affect the performance of the bonded joints, the stacking sequence playing an essential role with thin plates/tubes, mainly because of its remarkable effect on bending stiffness which, in turn, determines the deformation of the joint and subsequent failure. Matthews et al. [30] concluded that the non-linear analysis to predict the joint strength is necessary. Various types of failures encountered in FRP composite bonded joints are as follows:

- Tensile failure in the fiber direction
- Tensile failure perpendicular to the fiber direction
- Interlaminar shear failure
- Cohesive failure
- Interfacial failure

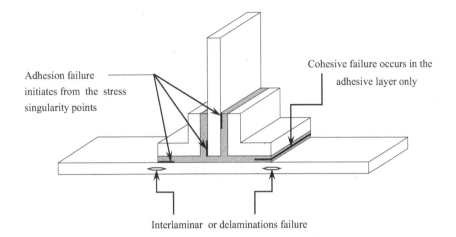

Cohesive failure occurs in the adhesive layer only

Adhesion failure initiates from the stress singularity points

Interlaminar or delaminations failure

**FIGURE 1.8**   Prominent failure modes in an adhesively bonded out-of-plane laminated fiber-reinforced polymeric composite joint.

However, because the laminated FRP composites have comparatively low interlaminar strength, the three prominent modes of failure in FRP composite out-of-plane bonded joints are as shown in Figure 1.8.

## 1.6   FUNCTIONALLY GRADED ADHESIVELY BONDED JOINTS

FGA represents a new class of non-homogeneous materials. The constituents gradually vary in some direction to impart macroscopically varying properties, often by design, to meet specific functional requirements. They possess variations in constituent volume fractions that lead to continuous change in the composition, microstructure, porosity, etc. This results in gradients in the mechanical and thermal properties.

FGAs are innovative composite materials whose composition and microstructure vary in space following a predetermined law. The gradual change in composition and microstructure gives a gradient of properties and performances. The schematic of composition and property variation is shown in Figure 1.9. From the viewpoint of mechanics, the distinguishing feature of FGA is that they are non-homogeneous not only concerning their thermo-mechanical properties but also in their strength-related properties such as yield strength, fracture toughness, fatigue and creep behavior.

Functionally graded materials (FGMs) appear in nature to reduce stress concentrations along with bi-material interfaces. Biological interfaces, such as dentin–enamel junction or tendon-to-bone [10], use the concept of FGMs. Nature offers many examples of graded materials to reduce stress concentrations along with the material interfaces. Tendon-to-bone joints are examples whose graded material properties allow more even stress distributions across the joint.

Today, this concept reduces stress concentrations appearing along the plate/tube-adhesive interfaces of the adhesive joints serving under static, dynamic and thermal loads. To relieve high-stress concentrations at the overlap region's free edges and have more uniform stress distributions, an adhesive layer with variable modulus has

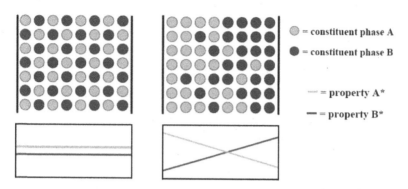

**FIGURE 1.9** Composition and property variation in conventional composite and functionally graded material.

been proposed. This requires at least the use of two adhesives with different mechanical and thermal properties as the adhesive layer. This is a primitive approach for the FGA layer. The concept of using multimodulus adhesives can provide improvements in the overall joint strength. This concept can also be implemented to the thermal stress problems of the adhesive joints to withstand low and high temperatures.

The use of more than one adhesive has been proposed to modify the mechanical properties of the adhesive along the overlap. This technique of using more than one adhesive consists of using a stiff and strong adhesive in the middle of the overlap and a flexible and ductile adhesive at the ends of the overlap to relieve the high-stress concentrations at the ends of the overlap. This allows a more uniform stress distribution, which leads to increased joint strength compared with a stiff adhesive alone. Although this approach has been discussed theoretically, there have been relatively few published experimental demonstrations of a practical method that yields significant improvements in the joint performance.

## 1.7 MANUFACTURING OF FUNCTIONALLY GRADED ADHESIVES: CHALLENGES AND PROSPECTS

FGAs can be defined as tailored adhesives with varying gradual mechanical properties along the desired dimension, allowing a more uniform stress distribution along the bond line. Joint improvement regarding joint strength has been achieved by manufacturing graded joints with microparticles, nanoparticles, tailored adhesives using adhesive mixes and graded cure processes [31].

### 1.7.1 MIXED ADHESIVE JOINTS

Mixed adhesive joints (MAJs) are a combination of stiff and ductile adhesives, resulting in a stepwise variation of properties along the adhesive bond line [32–35]. Combining a brittle adhesive at the center of the joint and ductile adhesive at the edges (Figure 1.10) provides a synergetic effect [33,36], allowing the load transfer from the overlap ends prone to stress concentrations to the middle higher strength

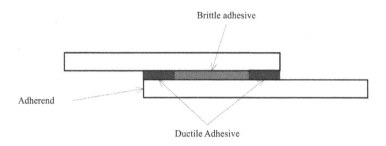

**FIGURE 1.10**   Mixed adhesive joint.

adhesive. Studies of these researchers target the combination of ductile and brittle adhesives, where the main concerns are the brittle-ductile adhesive quantities and corresponding lengths placed along with the joined substrates and their assembly during manufacturing.

Overall, bi-adhesive joint performance depends on the ductile-brittle adhesive ratio and position along the overlap. Brittle adhesives provide joint strength but do not take advantage of the whole overlap length, and on the flipside using too much ductile adhesive may lower joint strength. Nevertheless, the load is transferred along the overlap to the brittle adhesive with the right amount of ductile-brittle adhesive [33]. For an MAJ to be stronger than a standalone brittle or ductile adhesive joint, a higher section of the overlap has to work under load. This is achieved by the synergistic effects of the brittle-stiff adhesive combination, in which the load carried by the stiff adhesive has to be higher than the load carried by the compliant one [33].

The manufacturing of these joints is not straightforward because of the need to control both adhesives' length and thickness. Marques and da Silva [36] used a very stiff epoxy adhesive, Araldite® AV138/HV998 (Huntsman – Pamplona, Spain) and a more flexible epoxy adhesive, Araldite 2015 (Huntsman – Pamplona, Spain), where the brittle adhesive was placed over the lower patch, between the nylon fiber barriers and the ductile adhesive on the remaining parts. Both adhesives were cured simultaneously. Marques and da Silva [36] found that the most effective way to set the boundaries between the adhesives was to use a nylon line glued to the adherends, allowing excellent dimensional control at the expense of a small area compared with partitioning made using straps of Teflon or silicone. This technique is tricky manufacturing-wise because adding a glued thin nylon line adds another degree of complexity to the process.

### 1.7.2   ADHESIVE REINFORCEMENT USING SECOND-PHASE INCLUSIONS

The inclusion of a second phase to improve fracture toughness in epoxy polymers and adhesives has been a successful and proven method since the late 1960s [31]. This second phase may be a rubbery phase (reaction-induced phase separation) or reinforcement inclusion (fibers, whiskers, particles) [31,37]. Regarding the incorporation of a rubbery phase in epoxy polymers, authors in [37] used rubber particles to modify the epoxy matrices, such as liquid carboxyl-terminated butadiene-acrylonitrile (CTBN), methacrylate butadiene-styrene (MBS) copolymer particles and MBS with a few percent of carboxyl groups in a poly (methyl methacrylate) shell, which improved fatigue crack growth resistance for certain sizes. In this technique, the epoxy polymers are modified by adding CTBN or MBS rubber particles with varied volume fractions and particle sizes. Afterward, they were tested by performing fracture toughness tests on single-edge notched and compact tension specimens. The biggest concerns in manufacturing were preventing bubble formation in the specimens, where the resin was degassed before and after the addition of either the MBS particles or liquid CTBN rubber [31].

For nanoparticles and whiskers usage as reinforcing phases, a heat-resistant, polymer-based adhesive reinforced with carbon nanotubes and silicon carbide whiskers for low- and high-temperature applications, investigated by Wang et al. [38], displayed increased strengthening and toughening performance between room temperature and 700°C. Both these and carbon nanotubes were pretreated to prevent agglomeration; these pre-treatments included the mix of the said reinforcements in an ethanol solution of silane and several ultrasound washes.

Barbosa et al. studied the use of microcork particles to enhance fracture toughness of brittle adhesives [39–43]. This material was investigated due to the cork's several attractive properties such as good thermal, acoustic and vibration insulation, flexibility, substantial permeability to liquids and gases and impact absorption. Barbosa et al. found that microcork particles successfully acted as crack stoppers for specific volume concentrations and particle size ranges [39], improving the adhesive's maximum strain to failure lowering the composites' glass transition temperature ($Tg$). The Fourier-transform infrared spectroscopy analysis showed that the cork microparticles did not have any chemical reaction with the used adhesive [41]. This results in an overall increase in ductile behavior [40] and an increase in critical energy release rate ($G_{IC}$) [42].

Several researchers also used microencapsulated healing agents included in the resin, which promoted crack shielding and thus increased fatigue life extension and fracture toughness [43–45].

Fine-tuning of the distribution of micro- and nanoparticles along the bond line is one of the cornerstones required to obtain reliable FGA using these particles. For this effect, some important aspects can be listed, such as the chemical interaction between the added particles and adhesive, particle surface treatments to prevent agglomeration, optimal particle size and concentration, as well as what kind of particles act as the best reinforcement (particle nature) [31].

### 1.7.3 Functionally Graded Adhesive by Graded Cure

Carbas et al. [46] took a different approach and successfully manufactured functionally graded single-lap joints (SLJs) through a graded cure, obtaining higher joint performance than regularly cured SLJs. The graded cure was executed using a specially developed apparatus for SLJs composed of two heating coils, located near the overlap ends and a cooling coil situated at the middle of the overlap, thus generating a temperature gradient along the overlap (Figure 1.11). A positioning system was also built to guarantee the correct alignment, geometry, position and thickness of the adhesive layer.

Since this technique uses high-frequency electricity to heat materials, either the joint or the adherends need to have some kind of metallic properties, such as

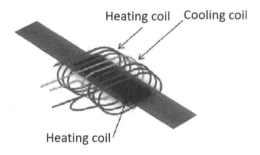

Heating coil  Cooling coil

Heating coil

**FIGURE 1.11** Functionally graded cure in single-lap joints [31].

ferromagnetic particles in the adhesive or metallic adherends [31]. When such particles are used with adhesives, the mechanical behavior is sensitive to the cure temperature for observing high cure temperature ductile behavior. For lower cure temperatures, a more brittle behavior occurs. The apparatus enables the production of an SLJ that has high adhesive stiffness at the center and minimum at the ends of the overlap. Following the same reasoning presented in the MAJ description, this leads to a higher joint strength than a joint made of the adhesives individually. An isothermal cure process was used to reference the graded cure results for comparison. Both adhesives used in this study were Araldite 2011 (Huntsman – Basel, Switzerland) and Loctite Hysol® 3422 (Henkel – Dublin, Ireland). The adhesives, when cured at high temperatures (100°C and 120°C), showed ductile behavior and, in contrast, when cured at low temperatures (23°C and 40°C), displayed brittle behavior.

In contrast to the isothermal-cured joints, the failure load of the functionally graded joints displayed an increase in failure load of 68% and 67% for Araldite 2011 cured at low and high temperatures, respectively. Functionally graded SLJs of Loctite Hysol 3422 showed a rise of 245.5% and 60.6% compared with their isothermally cured counter parts cured at low and high temperatures, respectively. Strain wise, both FGAs were better or just as good as their ductile non-graded counterpart was. Carbas et al. [46] also proved that an increase in adhesive thickness of the FGA SLJs was matched with a strength increase, which matches the behavior of ductile adhesives [47].

## 1.8 BONDED JOINT DESIGN PHILOSOPHY

The design of adhesively bonded joint must be based on (i) the nature of the materials to be joined, (ii) the configuration of the joint, that is, geometry and configuration, (iii) the joining methods and (iv) the strength and failure analysis. So, for a structural bonded joint, a designer should consider the following aspects:

- Service conditions such as stresses and environmental conditions are likely to be encountered in service.
- Selection of material combinations such as selection of suitable adhesive and FRP composite plates/tubes.
- Selection of specific joint configuration according to the need.
- Manufacturing specifications such as surface pretreatment and fabrication procedures.

Many theories for adhesively bonded joints have been developed for stress and strength analysis. Many authors have made various assumptions regarding the behavior of adhesive and plates/tubes in terms of differential equations. They have investigated the effects of multiple factors on the stress and strength. These factors are adhesive plasticity, large deformations and rotations, satisfaction of the boundary conditions at the overlap ends, spew fillet and geometry, bond line thickness, overlap length, etc. It has been shown that in addition to the large deformation, adhesive plasticity is another critical factor and cannot be ignored for the appropriate prediction of joint strength. While implementing the adhesive's non-linearity, the joint's analytical solutions become cumbersome. Analytical and closed-form solution analysis would be very difficult or maybe impossible when material (both adhesive and plates/tubes) non-linearity, delamination and debonding damages are considered. The mechanics of materials and fracture mechanics-based approach are considered the most suitable analysis tool for predicting strength of adhesively bonded joints.

It has been observed that the adhesively bonded joints often fail due to the damages initiated from the stress singularity locations and their propagation either in the adhesive layer or along the interfacial surfaces or in a combination of both. Thus, fracture mechanics has been widely used to correlate damage propagation behavior in adhesively bonded joints. The main governing parameters are SERPs for characterizing strength and service life and designing rationally adhesively bonded joints. The main governing parameters are SERPs. Figure 1.12 shows the three separate individual modes of damage propagation in the laminated FRP composite joints. These modes are (i) opening mode called Mode *I*, (ii) shearing mode known as Mode *II*, and (iii) tearing mode designated as Mode *III* [48].

A mixed-mode of delamination/interfacial damage in laminated FRP composite bonded joint is shown in Figure 1.13 [48]. A crack line damage in the bondline of adhesively bonded joint is constrained and will propagate under mixed-mode conditions in a direction such that the crack tips are in pure Mode *I* [49].

During the last few decades, numerous joint theories have been developed by many authors using a lot of simplifying assumptions concerning the behavior of the plate/tube and the adhesive of the bonded joints. These assumptions may enable

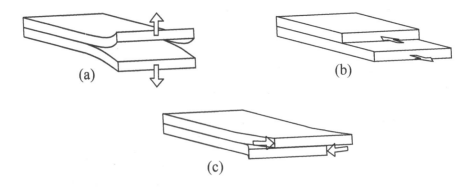

**FIGURE 1.12**   Three different individual modes: (a) Mode *I*, (b) Mode *II* and (c) Mode *III* of delamination/interfacial damage in laminated fiber-reinforced polymeric composite joint [48].

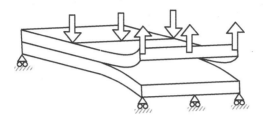

**FIGURE 1.13**   Mixed-mode delamination/interfacial damage in laminated fiber-reinforced polymeric composite joint [48].

removing the stress singularities that occur at the stress singularity locations. These stress singularities are due to dissimilar material properties at the interface, discontinuities of geometry, loading and material heterogeneity, etc. These are the main concerns for the joint designers. The details of assumptions made for stress analysis of bonded joints have been discussed in the pioneering work by Carpenter [49] for obtaining a closed-form solution. He concluded that the effect of a given assumption on predicted adhesive stress is challenging to determine from the differential equations.

### 1.8.1   Stress Analyses

Stress analysis is one of the critical steps for any structural design. It provides much vital information about the stresses and the strains in the real structures made of adhesively bonded joints subjected to specified loading and service conditions. This information will enable the designer to predict a bonded joint's strength and life. The stress analysis of bonded joints is a real challenge for two aspects of joint design: (i) because bi-material interfaces and geometric discontinuities create stress concentrations and material behavior uncertainties and (ii) the stress gives the idea of failure initiation in terms of failure index. Failure index refers to a parameter that characterizes the location of failure or damage initiation. The out-of-plane stresses are the most important factors responsible for the propagation of initiated damages. In the case of laminated FRP composite plates/tubes, interfacial/delamination-induced damages are the major threat to the bonded joint applications.

Stress analysis techniques can generally be classified into two major categories, that is, analytical solution, which is based on several mathematically simplified assumptions and numerical solutions using the FEA. Because of the complexities, analytical solutions of the bonded joint exist for simple geometry, loading and boundary conditions. Therefore, more emphasis is put on numerical methods when laminated FRP composites are used for plate/tube material and the joints involving complicated geometry, loading and boundary conditions. Among various numerical techniques available, it is seen that the FEA is not only simple and robust but also straightforward and versatile enough to cover all types of bonded joint problems relevant to practical situations. As described by Zienkiewicz [50], the FEA is a well-established numerical means for stress analysis of bonded joints. The FEA avoids the approximation of the closed-form theories.

# 2 Modeling and Damage Analyses of Functionally Graded Adhesively Bonded Joints in Laminated FRP Composites

## 2.1 INTRODUCTION

Damage/defect in a bonded structure is defined as any unintentional local variation in the physical state or mechanical properties, which may affect the component's structural behavior. This is often termed as a discontinuity or flaw in an intact structure. Damage calumniates the proper functioning of the structure, eventually causing its failure. Failure of a component or structure is defined as a state 'when a component or structure is unable to perform its primary functions adequately'.

Damages in adhesively bonded laminated FRP composites may manifest themselves in the forms of cohesive failure, interfacial failure, interlaminar/intralaminar delamination, debonding of fiber–matrix interface and fiber breakage. While all or some of the above damage modes may simultaneously be present in any structural bonded joint, damages due to the cohesive failure, interfacial failure and delamination are of prime concern because they may reduce the strength and stiffness of the structure significantly. So this has been a subject of major concern in engineering applications of the adhesively bonded joints because of the problems of structural integrity and stability, reduction in load-carrying capacity, stiffness reduction and exposure of the interior to an adverse environment, and final catastrophic failure of the structure. Delamination mode of failure in the laminated FRP composite plates/tubes is, especially of insidious nature. FRP composites exhibit relatively poor resistance to delamination damages. These damages may arise from micro-cracks and cavities or voids formed during manufacturing stages, service or maintenance induced damages or from low-velocity impact damage [15–18]. The structural degradation and stability reduction of composite structures with bonded joints are more critical due to the damages mentioned earlier.

Many researchers have investigated the influence of various parameters on the failure behavior in their studies related to composite bonded joints [51–54] experimentally and numerically for adhesively bonded composite joints. In those studies,

DOI: 10.1201/9781003201113-2

the typical bonding parameters are surface conditions (e.g., contamination, abrasion and plasma treatment), fillet, bond line thickness, surface ply angle, stacking sequence, environmental conditions etc. Many researchers have also predicted the onset of failures for the composite bonded joints [54–59]. But the failure prediction of the composite bonded joints is still difficult because the failure strength and failure mode differ according to the various bonding methods and parameters.

There have been usually two kinds of failure prediction methods discussed: one is stress or strain-based, and the other is based on the fracture mechanics approach. The stress- or strain-based method uses failure criterion equations, which include maximum stresses or strains in the bonded joint. This method is simple but may not be suitable because there are locations of stress or strain singularities in the bonded joints. In the fracture mechanics method, an initial crack is assumed and crack growth is assessed by comparing the computed strain energy release rate (SERR) with fracture toughness determined by experiments. The fracture toughness differs according to the mode mix ratios and failure modes (for example, interfacial, cohesive and delamination failure). So, the fracture toughness test is a time (and cost) consuming process. In addition, the fracture mechanics method is not suitable for the bonded joint, which fails without an initial crack. As mentioned above, these two methods have merits and demerits, and various failure modes may appear due to the mechanical properties of adhesives or bonding methods in the composite bonded joints.

In recent decades, researchers have been investigating new material system developments and their applications to existing engineering systems and their failure analysis to improve these systems.

Figure 2.1a and b describes the crack tip and crack front configurations for studying straight-edged and curved interlaminar fracture initiation and progression. Delamination in a composite material is fundamentally an interlaminar fracture phenomenon involving debonding or separating two highly anisotropic, fiber-reinforced laminates. In many types of composite structures (e.g., aircraft, marine), delamination is the most common form of defect/damage. In addition, sometimes, manufacturing deficiencies cause inadequate bonding between the layers. This may lead to delamination over a long period in service. Several mechanisms contribute to property degradation, and they have received considerable attention in recent

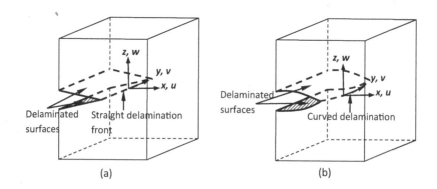

**FIGURE 2.1**   Schematic of delamination fronts: (a) straight and (b) curved [48].

times as they differ remarkably from the failure process observed in conventional metallic components. Fracture mechanics has been extensively used for damage analysis and prediction for isotropic materials. As such, the application of linear elastic fracture mechanics has gained the attention of the researchers for investigating the delamination mode of failure pertaining to the adhesively bonded joints of laminated FRP composites.

## 2.2 MECHANICS OF INTERFACIAL/DELAMINATION DAMAGES IN ADHESIVELY BONDED JOINTS OF LAMINATED FIBER-REINFORCED POLYMERIC COMPOSITES

In this section, failure modes of the adhesion/interfacial failure, which usually initiates at the stress singularity points and propagates along the interfacial surfaces of the out-of-plane joints, have been addressed. The joints are embedded with adhesion/interfacial failures emanating from the critical locations and subjected to different loadings, leading to mixed-mode failure conditions. Some typical failure modes and their propagation for single and double supported tee joint, tubular lap joint and tubular socket joint are shown in Figure 2.2.

The other failure modes in adhesively bonded joints are due to the failure of laminated composite plates/tubes. Due to the laminated nature of the FRP composite members and the relative weakness in the through-the-thickness direction, the failure mechanism of adhesively bonded joints has become the focus of research. It has been observed that the magnitudes of the failure indices at the adhesive midlayer are comparatively smaller compared with that of the plate/tube-adhesive interfaces. This indicates that bonded out-of-plane joints are more prone to interfacial failures than cohesive failures.

Figure 2.2b shows cohesive failure confined to the adhesive layer in a single supported tee joint. The possible adhesion/interfacial failure initiation is represented in Figure 2.2c for the double supported tee joint. Failure due to delamination damages in adhesively bonded out-of-plane joints is also shown in Figure 2.2f, which indicates that tubular socket joints may fail due to the propagation of delamination damages in the FRP composite tubes.

## 2.3 METHODOLOGY FOR DESIGN AND ANALYSIS OF FUNCTIONALLY GRADED ADHESIVELY BONDED OUT-OF-PLANE JOINTS

Three-dimensional stress analysis plays a vital role in any structural design. It provides many important information about the stresses and the strains in the actual structures made of adhesively bonded out-of-plane joints subjected to specified loading and service conditions. This information will enable the researchers to predict the strength and service life of the designed structures. Accurate determination of three-dimensional stress distribution in adhesively bonded out-of-plane joints represents one of the most challenging structural stress analysis problems. Bi-material interfaces and geometric discontinuities create stress concentrations

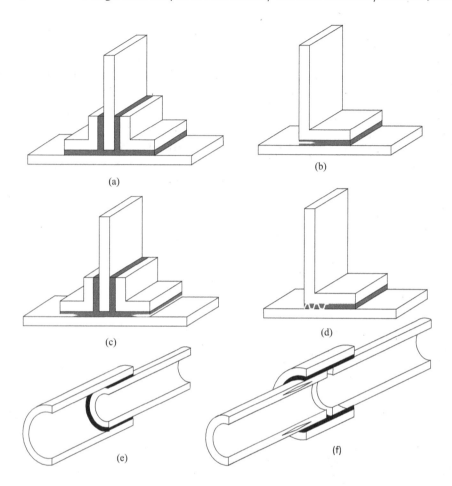

**FIGURE 2.2** Different types of failure modes and their propagation paths in adhesively bonded out-of-plane joints: (a) double supported tee joint without damage, (b) cohesion failure propagation path in single supported tee joint, (c) possible interfacial failure initiation and propagation path in double supported tee joint, (d) wavy failure propagation path along the two interfacial surfaces in single supported tee joint, (e) interfacial failure propagation path in tubular lap joint and (f) delamination damages in tubular socket joint.

and uncertainty in material behaviors arises. Because of the complexities, analytical solutions of the bonded joint exist for simple geometry, loading and boundary conditions. Therefore, more emphasis is put on numerical methods when laminated FRP composite plates/tubes are used as adherend and functionally graded material as adhesive and the joints involving complicated geometry, loading and boundary conditions. Among various numerical techniques available, it is seen that the FEA is not only simple and robust but also straight forward and versatile enough to cover all types of out-of-plane functionally graded adhesively bonded joint problems relevant to practical situations.

Accurate prediction of strength and service life of adhesively bonded out-of-plane joints has been one of the challenging tasks for assessing the structural integrity of any bonded structures. There is a lack of complete understanding of failure mechanisms in various types of adhesively bonded out-of-plane joints. The joint failure mechanism includes the onset of initial failure, stable failure propagation, and catastrophic damage propagation. This phenomenon becomes complex and much involved when the plates/tubes of out-of-plane joints are made from laminated FRP composites, and adhesive is functionally graded material. There are many failure criteria available that have been used to predict the onset of damages in the adhesively bonded joints. Two approaches have been used to indicate the location of the onset of damages. One of them is based on the mechanics of materials approach, and the other follows the fracture mechanics procedures.

### 2.3.1 CRITERIA FOR DAMAGE ONSET IN OUT-OF-PLANE BONDED JOINTS

One of the major concerns in laminated FRP composite out-of-plane joint is the prediction of location of damage initiation due to the prevailing tri-axial states of stresses which three-dimensional FEA has accurately evaluated. Adhesively bonded out-of-plane joint experiences two important modes of mechanical failure: (i) interfacial failure, also known as adhesion failure, which occurs at the interface of adhesive and plate/tube, and (ii) cohesive failure within the adhesive layer. The interfacial failure initiates from the stress singularity points of adhesively bonded out-of-plane joint structures. Under three-dimensional stress states, the onset of failures over the bond line interfacial surfaces has been predicted by Tsai and Wu [60] coupled stress failure criterion, which takes into account the interaction of all six stress components, which is given by

$$
\frac{\sigma_x^2}{X_T^2} + \frac{\sigma_y^2}{Y_T^2} + \frac{\sigma_z^2}{Z_T^2} + \frac{\tau_{xy}^2}{S_{xy}^2} + \frac{\tau_{yz}^2}{S_{yz}^2} + \frac{\tau_{xz}^2}{S_{xz}^2} + \sigma_x\left(\frac{1}{X_T} - \frac{1}{X_C}\right) + \sigma_y\left(\frac{1}{Y_T} - \frac{1}{Y_C}\right) +
$$

$$
\sigma_z\left(\frac{1}{Z_T} - \frac{1}{Z_C}\right) + f_{xy}\sigma_x\sigma_y + f_{yz}\sigma_y\sigma_z + f_{xz}\sigma_x\sigma_z = e^2 \begin{cases} e \geq 1, & \text{failure} \\ e < 1, & \text{no failure} \end{cases}
$$

(2.1)

where $X_T$, $Y_T$ and $Z_T$ are the allowable tensile strengths in the three principal material directions, $X_C$, $Y_C$ and $Z_C$ are the allowable compressive strengths in the three principal material directions, and $S_{xy}$, $S_{yz}$ and $S_{xz}$ are the shearing strengths of the orthotropic layer in various coupling modes. The coupling coefficients reflecting the interaction between $x$, $y$ and $z$ directions are given by $f_{xy}$, $f_{yz}$ and $f_{xz}$, respectively. Failure index ($e$) is defined as the parameter to evaluate the condition whether the bonded joint is likely to fail or not. If $e \geq 1$ failure occurs, else there is no failure.

Interfacial/adhesion failure is considered as a delamination damage which is mainly attributed to the interlaminar stress effects, so only the interlaminar shear stresses ($\tau_{xz}$ and $\tau_{yz}$) and through-the-thickness normal stress ($\sigma_z$) are required to be used to predict the damage initiation in terms of failure index values '$e$'. Therefore, the Tsai-Wu criterion as given in Eq. (2.1) can be simplified as:

i. Interfacial failure in tension, for $\sigma_z > 0$;

$$\left(\frac{\sigma_z}{Z_T}\right)^2 + \left(\frac{\tau_{xz}}{S_{xz}}\right)^2 + \left(\frac{\tau_{yz}}{S_{yz}}\right)^2 = e^2 \begin{cases} e \geq 1, & \text{failure} \\ e < 1, & \text{no failure} \end{cases} \qquad (2.2)$$

ii. Interfacial failure in compression, for $\sigma_z < 0$;

$$\left(\frac{\sigma_z}{Z_C}\right)^2 + \left(\frac{\tau_{xz}}{S_{xz}}\right)^2 + \left(\frac{\tau_{yz}}{S_{yz}}\right)^2 = e^2 \begin{cases} e \geq 1, & \text{failure} \\ e < 1, & \text{no failure} \end{cases} \qquad (2.3)$$

where $Z$ is the interlaminar normal strength and $S_{yz}$ and $S_{xz}$ are the interlaminar shear strengths, respectively, among the two orthogonal shear coupling directions. Because of material symmetry, the FRP composite laminates considered in this research have $S_{yz} = S_{xz}$.

Failure criteria developed for isotropic materials have been used to predict the cohesive failure in the adhesive layer of adhesively bonded structures. Raghava et al. [61] have proposed a parabolic yield criterion as given below, and this has been used for the prediction of the location of cohesive failure initiation in an adhesively bonded out-of-plane joint

$$(\sigma_1 - \sigma_2)^2 + (\sigma_2 - \sigma_3)^2 + (\sigma_3 - \sigma_1)^2 + 2(|Y_C| - Y_T)(\sigma_1 + \sigma_2 + \sigma_3)$$

$$= 2e|Y_C|Y_T \begin{cases} e \geq 1, & \text{failure} \\ e < 1, & \text{no failure} \end{cases} \qquad (2.4)$$

where $\sigma_1$, $\sigma_2$ and $\sigma_3$ are the allowable principal stresses causing yield and $|Y_C|Y_T$ are the absolute values of the compressive and tensile yield stresses, commonly known as strengths. It may be noted that when $|Y_C|$ both strength values in tension and $Y_T$ compression are equal, the above yield criterion reduces to the most familiar von Mises cylindrical criterion.

## 2.3.2   DAMAGE GROWTH IN OUT-OF-PLANE BONDED JOINTS WITH STRAIN ENERGY RELEASE RATE

To properly understand stable and unstable crack growth at different crack lengths, an evaluation of energy release rates is necessary. The mixed-mode critical energy release rate ratio indicates stable continuous crack growth through the laminate specimen [62]. SERR formulae were predominantly used to characterize the interlaminar fracture in delamination and debonding analysis of composite structural members [24,63–68]. Calculations of energy release rate components are based on

Irwin's theory of the crack closure technique [69]. The energy release rate procedure is robust as it is based on a sound energy balance principle and mode separation of SERR is possible. This can be easily incorporated into an analytical formulation and compared with critical fracture energy values to efficiently predict interfacial failure growth and propagation.

The behavior of an existing pre-embedded interfacial failure/delamination is governed by the values of three modes of SERR around the delamination front. However, in composite laminates, exact closed-form expressions for the energy release rates are not possible due to their inherent complications. This leads to the finite element (FE) evaluation of energy release rates based on linear elastic fracture mechanics principles. Estimation of the SERRs $G_I$, $G_{II}$ and $G_{III}$ is the central notion of an interfacial failure propagation analysis. This interlaminar fracture energy release rate calculations depend primarily on the nature of the applied stress state and the crack geometry. Analytical solutions are available only for some simple stress states and geometries. However, the practical utility of such straightforward analytical solutions is limited. Implementing the fracture mechanics approach relies on estimation (typically by analytical or numerical techniques) of the applied SERR. The behavior of the straight free-edge problem can be analyzed as a two-dimensional problem. However, the behavior around a curved free edge or corner is inherently a three-dimensional problem. The energy release rate, denoted by $G$ is defined for virtual crack extension in the same plane as

$$G = \frac{dW}{da} - \frac{dU}{da} \qquad (2.5)$$

where $W$ is the work done by the external traction per unit thickness, $U$ is the strain energy of the body per unit thickness, and $a$ is the crack length.

The physical significance of the energy release rate corresponds to the rate of change of energy of the system per unit area of crack growth. This value is compared with the toughness parameter $G_c$ (material tests) for studies on delamination onset and growth. In simplifying both analysis and test requirements, it is convenient to decompose the applied SERR and the toughness into parameters corresponding to three possible cracking modes. Damage may be stressed in three different modes, these modes acting independently or together (Figures 1.12 and 1.13). Three SERRs ($G_I$, $G_{II}$ and $G_{III}$) and three material property parameters ($G_{IC}$, $G_{IIC}$ and $G_{IIIC}$) are defined as corresponding to fracture $I$, $II$ and $III$ modes, that is, opening, shearing and tearing, respectively. The three critical SERR parameters $G_{IC}$, $G_{IIC}$ and $G_{IIIC}$ can be related to three fracture toughness parameters $K_{IC}$, $K_{IIC}$ and $K_{IIIC}$. Thus, after evaluation of $G_I$, $G_{II}$ and $G_{III}$, a comprehensive idea of the structure's structural integrity is available.

### 2.3.3 FINITE ELEMENT MODELING AND SIMULATION OF INTERFACIAL FAILURE AND ITS PROPAGATION

For most general geometries and loadings encountered in practice, numerical techniques commonly calculate the SERRs. FE method is the most widely accepted procedure followed in the numerical analysis [68]. The advantage of the FE method is that it imposes no inherent restriction on geometry, applied stress state and material behavior. The approach to estimating SERRs using FE methods is initially to evaluate the $(W - U)$ term in Eq. (2.5) for two models, one with an existing crack and the second considering an incremental growth in crack surface area. Although this approach is helpful in some instances but possesses difficulty in general applications. The main problem is that a $G$ value for the structure is obtained with no information on its distribution along the crack front nor on the relative magnitude of the individual modes. Hence, except in some specific cases, it is impossible to apply the results in a failure criterion requiring a direct comparison of SERR with material toughness. An alternative approach, still in reasonably everyday use, is to calculate $G$ by integrating energy along a contour around the crack front, that is, determination of the $J$-integral. This has the benefit of estimating a local SERR and evaluation for each model. However, this requires the use of special collapsed elements at the crack front, and this can add significantly to mesh and hence model complexity. The further problem is that it is not strictly applicable to cases where dissimilar materials on opposing sides of the crack tip give rise to an oscillating singularity, requiring a three-dimensional analysis. In response to the above difficulties, several alternative approaches have been developed to apply the FE method. One of the most useful is the virtual crack closure technique. It is also sometimes addressed in literature as modified crack closure integral (MCCI).

### 2.3.3.1  Modified Crack Closure Integral

The curvature plane and normal are constant everywhere for a straight-edged crack front. So mode definition is intuitive and constant for the entire front. Individual modes are well defined along the crack front. Mode *I* is caused by the out-of-plane crack opening, Mode *II* by the shear perpendicular to the straight delamination/crack front and mode *III* by the shear component tangential to the front. The advantage is that a single model can estimate the SERRs along the delamination/crack front. A major benefit of the method is that it allows accurate estimation of SERR while using a relatively simple mesh (although a fine mesh may still be needed at the crack tip).

Furthermore, it allows the separation of individual modes of SERRs $G_I$, $G_{II}$ and $G_{III}$ by isolating the relevant stresses and displacements for inclusion in the integrations. Furthermore, benefits include (i) the calculation of these values on an element-by-element basis and provision of results on a node-by-node basis and (ii) straightforward treatments of oscillatory singularities in stress. However, this makes some assumptions regarding possible changes in the stress state for an incremental closure of the crack. Numerical analyses evaluate the stress and displacement fields ahead of the damage front. Figure 2.3 shows the self-similar damage front propagation schematic of $\Delta a$ crack length.

Consider a 3D FE idealization with FEs symmetric about the damage front as shown in Figure 2.4.

Suppose the interfacial failure propagates from $a$ to $(a + \Delta a)$, for small infinitesimal values of $\Delta a$. In that case, the opening displacement behind the new damage front will be approximately the same as that behind the original one. Then the work required for propagation of damage length from $a$ to $(a + \Delta a)$ is the same as that

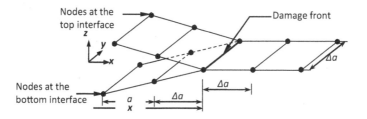

**FIGURE 2.3** Schematic of propagation of damage front in 3D analyses.

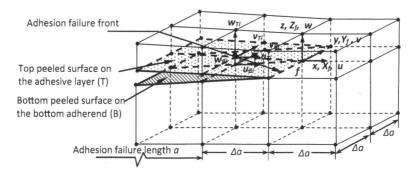

**FIGURE 2.4** Modified crack closure integral applied to the tee joint for propagation of adhesion/interfacial failure length 'a' at the interface between the bottom adherend (base plate) and adhesive layer.

necessary to close the virtually opened damage length $(a + \Delta a)$ to damage length $a$. Irwin [69] computed this work as

$$W = \frac{1}{2} \int_0^{\Delta a} \left[ \sigma_z (\Delta a - x, 0) \times w(x, 0) + \tau_{xz} (\Delta a - x, 0) \times u(x, 0) + \tau_{yz} (\Delta a - x, 0) \times v(x, 0) \right]$$

$$\times dx \tag{2.6}$$

where $w(x, 0)$, $u(x, 0)$ and $v(x, 0)$ are the damage opening displacements in modes $I$, $II$ and $III$ at a distance $(\Delta a - x)$ behind the damage front and $\sigma_z(\Delta a - x, 0)$, $\tau_{zx}(\Delta a - x, 0)$ and $\tau_{yz}(\Delta a - x, 0)$ are the corresponding stress components.

Irwin [69] conceptualized that the stress field ahead of the crack due to an infinitesimal small crack extension could be made equal to that behind it due to the crack closure, when the crack was assumed to propagate in the same plane. Rybicki and Kanninen [22] discussed the advantage of the method of MCCI based on the postulates of Irwin's theory of crack field solution for calculating the individual modes of SERRs for cracks in the bi-material interface. The SERR $(G)$ is given by

$$G = \lim_{\Delta a \to 0} \frac{W}{\Delta A} = G_I + G_{II} + G_{III} \tag{2.7}$$

Another advantage of this MCCI method is that it is not very sensitive to mesh design [70,71]. Using the MCCI, the three individual modes of energy release rates can be expressed as follows:

$$G_I = \lim_{\Delta a \to 0} \frac{1}{2\Delta A} \int_a^{a+\Delta a} \int_{-\Delta a/2}^{\Delta a/2} \sigma_z(x,y) \times [w_T(x - \Delta a, y) - w_B(x - \Delta a, y)] \, dx \, dy \quad (2.8)$$

$$G_{II} = \lim_{\Delta a \to 0} \frac{1}{2\Delta A} \int_a^{a+\Delta a} \int_{-\Delta a/2}^{\Delta a/2} \tau_{xz}(x,y) \times [u_T(x - \Delta a, y) - u_B(x - \Delta a, y)] \, dx \, dy \quad (2.9)$$

$$G_{III} = \lim_{\Delta a \to 0} \frac{1}{2\Delta A} \int_a^{a+\Delta a} \int_{-\Delta a/2}^{\Delta a/2} \tau_{yz}(x,y) \times [v_T(x - \Delta a, y) - v_B(x - \Delta a, y)] \, dx \, dy \quad (2.10)$$

where $T$ and $B$ represent the top and bottom sublaminate parameters, respectively, and $a$ is the existing crack length, $\Delta a$ is virtual crack extension length due to the external loads, and $\Delta a$ is also the crack's width shown in Figures 2.3 and 2.4. The displacement parameters $[u_T, v_T, w_T]$ and $[u_B, v_B, w_B]$ are the corresponding values for the nodes at the top and bottom interface, respectively, just behind the propagated crack front, and $\sigma_z$, $\tau_{zx}$ and $\tau_{zy}$ are the stresses required to close the progressive crack front. Although MCCI has the advantage of mode separation of SERRs, it requires specific meshing techniques. The practical difficulties associated with the need for a very fine mesh include the following:

- Models can be time-consuming to generate.
- Analysis tends to be highly computationally intensive.
- Convergence studies are usually needed to verify that the mesh is adequate, and these are in themselves time-consuming.
- Proper understanding of debonding characteristics at the interlaminar region is essential in any interfacial damage analysis. The interfacial damage analyses of different types of out-of-plane bonded joints with mono-modulus and functionally graded adhesive have been carried out as a part of the ongoing research work. It has been observed that the bonding phenomena at the interface between plates/tubes and adhesive layer affect the structural integrity and load-carrying capacity of bonded joints of laminated fiber-reinforced polymeric composites severely.

### 2.3.3.2 Virtual Crack Closure Technique

In Rybicki and Kanninen's [22] method of virtual crack closure technique implementation in the FE analysis, the stress-based Eqs. (2.8)–(2.10) take the form of simple multiplication of corresponding nodal forces at the FE nodes on the damaged plane ahead of the failure front and the damage face displacements behind the damage front.

Because of this, the FE mesh needs to satisfy certain conditions [21,22]. These are (i) the FE mesh should be symmetric on the damaged plane as well as about the damage front, (ii) the element size $\Delta a$ at the damage front should be small and (iii) the normality of the FE mesh should be maintained near the damage front.

For the three-dimensional specimen, the damage length '$a$' is represented by two adjacent surfaces with node configuration, as shown in Figure 2.4. The nodes at the top and bottom surfaces of the discontinuities have identical coordinates. Suitable gap elements connect them to prevent interpenetration of the damaged surfaces. The damaged front and subsequent undamaged portions are modeled using pairs of nodes with identical coordinates coupled through multi-point constraint (MPC) elements to simulate the propagation of adhesion failure. The MPC elements along the damage front are activated during the FE simulations for self-similar failure propagation. The other MPC elements ahead of the damage front are passive. These passive MPC elements are activated in turn by incremental crack front displacement $\Delta$ to simulate the damage front propagation.

The mode $I$, mode $II$ and mode $III$ components of SERR, that is, $G_I$, $G_{II}$ and $G_{III}$, are computed accordingly and are given by

$$G_I = \frac{1}{2\Delta A} Z_f \left( w_{Ti} - w_{Bi} \right) \qquad (2.11)$$

$$G_{II} = \frac{1}{2\Delta A} X_f \left( u_{Ti} - u_{Bi} \right) \qquad (2.12)$$

$$G_{III} = \frac{1}{2\Delta A} Y_f \left( v_{Ti} - v_{Bi} \right) \qquad (2.13)$$

where $\Delta A = \Delta a \times \Delta a$ is the area virtually closed. Referring to Figure 2.4, $Z_f$, $_{If}$ and $X_f$, the opening, tearing and sliding mode forces, respectively, are required to hold the nodes at the damage front (at the node $f$) together to prevent it from opening and its subsequent propagation. The corresponding displacements behind the damage front at the top peeled surface of the adhesive layer (T) nodes are denoted by $u_{Ti}$, $v_{Ti}$ and $w_{Ti}$ and at the bottom peeled surface on the bottom adherend (B) nodes represented by $u_{Bi}$, $v_{Bi}$ and $w_{Bi}$. All forces and displacements are obtained from the FE analysis with respect to the global coordinate system.

The MPC element at the damage front helps to determine the values of the forces. The gap elements used behind the damage front give the values of the displacements of the upper and lower surfaces created due to the adhesion failure. Thus, $G_I$ and $G_{II}$ at any point on the damage front are determined by releasing the MPC elements to simulate the damage propagation.

# 3 Stress Analysis of Functionally Graded Adhesively Bonded Single Supported Tee Joint

## 3.1 INTRODUCTION

With the increasing use of adhesive bonding joints in industry, the analysis and design of the joints are more and more critical. In structural adhesive joints, the tee joint is one of the typical joint geometries used to bond two plates at a right angle or some other angle. The primary function of this out-of-plane joint is to transmit flexural, tensile and shear loads between two sets of panels meeting at the joint by an angle connection. For adhesively bonded tee joints, the structural performances are primarily dependent on various factors such as joint configuration, joint materials, service conditions and joining processes.

The expanding application of fiber-reinforced polymeric (FRP) composite materials for joints in the aerospace industry has extended to other sectors such as marine, automotive, chemical and defense industries. Tee joints become necessary in the structure of ships and boats made from FRP materials because large and complex structures cannot be formed in one process. T-stiffeners are used extensively in aircraft wings to prevent skin buckling during wing loading.

Although in-plane joints have received the most attention, different types of adhesively bonded out-of-plane joints have been used in many structural applications. Shenoi and Violette [72] analyzed tee joint geometry under out-of-plane loading for hull bulkhead structures using experimental and numerical methods. Li et al. [73] performed a two-dimensional stress analysis of a single supported tee joint in which a right-angled plate was bonded to a rigid plate with an adhesive using the finite element (FE) method. They have investigated stresses within the considered joint with varied bond lengths, adhesive and adherend thicknesses. It was noticed that these design variables significantly affect the stress distribution. The same authors in [74] also gave attention to the analysis of the stiffness of a single supported tee joint under varied loadings and recommended suitable dimensions of the tee joint. Panigrahi and Zhang [75] carried out geometrically nonlinear finite element analyses (FEAs) of an adhesively bonded single supported tee joint made of graphite/epoxy composite plates. Normal, shear and von-Mises stress distributions at the midsurface of the adhesive

DOI: 10.1201/9781003201113-3

layer have been evaluated. The authors noticed critical location for failure initiation in the adhesive layer around the corners of the right-angled plate. Apalak [76] showed that stress and deformation states of joint members were dependent on loading conditions. The same author also noticed that analysis using small strain–small displacement theory could be misleading to predict stresses and deformations for tee joints under large displacement. Apalak et al. [77] conducted FEA for bonded tee joints with single support with angle reinforcement for different loadings and boundary conditions. The plates and supports were made of steel, and adhesive was assumed to be of a linearly elastic material. In bonded joints, joint failure is expected to initiate in the adhesive region subjected to high-stress concentrations; therefore, peak adhesive stresses were evaluated in these critical regions. Apalak [76,78–80] also showed that geometrical nonlinearity had a significant effect on the stress distributions in the plates and adhesive layers of different types of single supported tee joints/corner joints.

Many approaches have been attempted to decrease high-stress concentration commonly occurring around the adhesive layer's free ends for any geometry of adhesively bonded joints. These include altering the adherend geometry [81,82], adhesive geometry [83] and spew geometry [84,85] and ply dropping technique. Harris and Adams [85] have performed nonlinear FEA with adherend made of elastic material, and adhesive was considered elastic-plastic. The authors evaluated the strength of the in-plane joint with a spew fillet. Their research indicated that rounding the corners at the overlapping end reduces the stress singularity. Sancaktar and Kumar [86] used rubber toughening in epoxy to increase joint strength. They have used FEA and experimental methods to validate the output of their research work. Some of the earliest works on the grading of the adhesive were reported by Patrick [87] and Raphael [88], where grading was discretely achieved using two adhesive materials. The main objective of Carbas et al. [89] was to study a functionally modified adhesive to have mechanical properties that vary gradually along the overlap of the joint, which results in uniform stress distribution along the overlap. It was found that joint strength with functionally graded adhesive was higher than that with mono-modulus adhesive. Recently, Breton et al. [90] determined the optimum grading of properties in a bond line to maximize the ultimate load capacity of a single-lap aluminum/composite joint under shear forces. The study is performed considering a grading strategy based on mixtures of compatible adhesives with dissimilar characteristics to obtain the desired variations in properties. The analysis is undertaken using the FE method as the calculation tool, taking into account the nonlinear behaviors of the structural adhesives selected for the study.

Ascribed to the published literature, research on adhesively bonded tee joints using an FGA adhesive layer made of FRP composite materials is limited to date. This chapter presents a numerical simulation of stress and failure analyses for the considered tee joint made of FRP composite plates using geometrically nonlinear FEA. FE model of the tee joint has been validated with the published results available in the literature [73]. A convergence study has been made to keep the error within 1% for the validated model. The out-of-plane and von-Mises stress components are evaluated on the mid-surface of the adhesive layer of the tee joint for three different FRP composites such as Gr/E, Gl/E, B/E with varied anisotropic ratio fiber orientation angle, viz. $[0]_8$, $[(0/90)_s]_2$, $[(+45/-45)_s]_2$. The effects of material anisotropy and laminate stacking sequence on out-of-plane and von-Mises stress components have been studied and accordingly, suitable design recommendations have been made. Furthermore, an attempt has been made to reduce the out-of-plane stress level at the ends of overlap for the tee joint using a functionally graded adhesive material. The bond line has been graded by two material gradation function profiles: linear and exponential. A systematic parametric study has been conducted with various modulus ratios to investigate their effect on out-of-plane and von-Mises stress distribution along the bond length.

## 3.2 MODELING AND ANALYSIS OF TEE JOINT

The three-dimensional state of stress and failure prediction of adhesively bonded out-of-plane joints are difficult to deal with analytically. An adequate solution need to account for geometrical discontinuities, out-of-plane loadings and material property variations, anisotropy and laminated plates' construction, functionally graded adhesive, etc. The linear elasticity theory cannot predict these structures' deformation and stress states correctly because it ignores the squares and products of partial derivatives of the displacement components concerning material coordinates. When these derivatives are not small, these terms result in nonlinear effect, that is, geometrical nonlinearity. It can be encountered in adhesive joints with an unbalanced geometry or subjected to loading and boundary conditions causing large displacements and rotations. Thus, three-dimensional FEAs with geometric nonlinearities have been performed to provide detailed insight into the design and failure of the single supported tee joint.

As shown in Figure 3.1, the tee joint consists of a right-angled plate adhesively bonded to a base plate [73]. Dimensions of the considered tee joint are shown in Table 3.1.

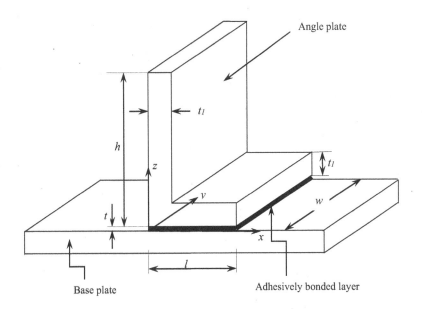

**FIGURE 3.1** Geometry and configuration of the tee joint made of fiber-reinforced polymeric composites.

**TABLE 3.1**
**Dimensions of Tee Joint [75]**

| Parameters | Dimensions (mm) |
| --- | --- |
| Bond length, $l$ | 15 |
| Plate thickness, $t_l$ | 2 |
| Plate height, $h$ | 40 |
| Adherend width, $w$ | 20 |
| Adhesive thickness, $t$ | 0.2 |

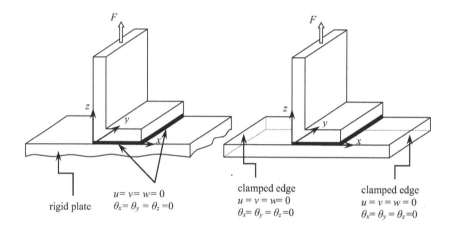

**FIGURE 3.2** Boundary conditions of tee joint with single support: (a) rigid base and (b) flexible base.

The joint is subjected to an out-of-plane load '$F$' of 50 N in $z$-direction applied through the right-angled plate. The considered tee joint is analyzed for two boundary conditions, that is, by assuming rigid and flexible base as shown in Figure 3.2.

Both the plates of the joint are made of FRP composite materials such as Gr/E, Gl/E and B/E, possessing orthotropic material properties as given in Table 3.2.

The solid elements called SOLID 46 and SOLID 45 of ANSYS [92] are used to model the FRP composite plates and adhesive layer of the tee joint, respectively. One element is used to model the composite plate, and two elements are used to model the adhesive layer through the thickness.

Solid 45 and solid 46, shown in Figure 3.3a and b, are three-dimensional isoparametric solid elements. SOLID 46 is a layered version of SOLID 45 with eight nodes, and each node possesses three translational degrees of freedom. These elements are accurate, suitable or appropriate for modeling adhesively bonded joints of laminated composites. To increase the computational efficiency graded meshing technique is used. Precisely very fine mesh is used to model the regions where high-stress gradients are expected. The regions with different discretization levels and the zoomed views are schematically depicted in Figures 3.4 and 3.5.

## TABLE 3.2
## Orthotropic Properties of Fiber-Reinforced Polymeric Composite Lamina [27,29,91]

| Elastic Properties | Graphite/Epoxy | Glass/Epoxy | Boron/Epoxy |
|---|---|---|---|
| $E_x$ (GPa) | 127.50 | 38.60 | 207.00 |
| $E_y$ (GPa) | 9.00 | 8.27 | 18.63 |
| $E_z$ (GPa) | 4.80 | 8.27 | 18.63 |
| $G_{xy} = G_{xz}$ (GPa) | 4.80 | 4.14 | 4.50 |
| $G_{yz}$ (GPa) | 2.55 | 4.00 | 3.45 |
| $v_{xy} = v_{xz}$ | 0.28 | 0.25 | 0.27 |
| $v_{yz}$ | 0.41 | 0.27 | 0.35 |

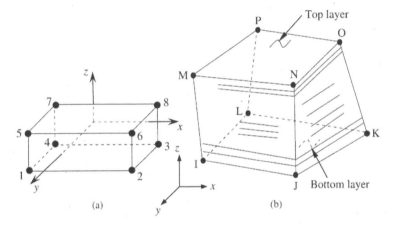

**FIGURE 3.3**  Various finite elements used for modeling the tee joint: (a) SOLID 45 and (b) SOLID 46.

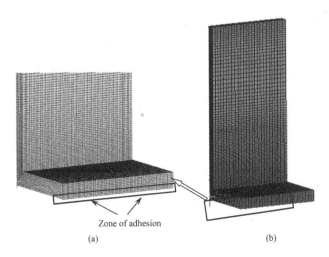

**FIGURE 3.4**  Discretized tee joint with rigid support: (a) zoomed view and (b) full model.

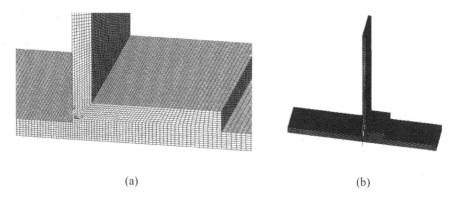

(a)                                                          (b)

**FIGURE 3.5**　Discretized tee joint with flexible support: (a) zoomed view and (b) full model.

---

**TABLE 3.3**
**Elastic Properties of Epoxy Adhesive [27,29]**

| | |
|---|---|
| $E$ (GPa) | 2.8 |
| $\upsilon$ | 0.4 |

---

This research considers two sets of adhesive material properties as given in Table 3.3 for mono-modulus adhesive and other with functionally graded material.

The continuous variation of elastic modulus of adhesive along the bond length has been considered. The bond length consists of flexible adhesive on the left part of the joint and stiffer adhesive on the right part of the joint. The elastic moduli of adhesive along the bond line are evaluated by two material gradation function profiles, that is, linear and exponential, as shown in Figures 3.6 and 3.7.

It was assumed that the relationship between Young's modulus and shear modulus remains constant invariably with grading or, in other words, Poisson's ratio '$\upsilon$' remains constant [7]. The linear [93] and exponential [94,95] variation of Young's modulus ($E$) is expressed as

$$E(x) = E_1 + (E_2 - E_1) \times \frac{x}{l} \quad \text{for linear function profile} \tag{3.1}$$

and

$$E(x) = E_1 \times \exp(\lambda x) \quad \text{for exponential function profile} \tag{3.2}$$

where '$\lambda$' is the material nonhomogeneity parameter. The value of $\lambda$ is expressed as

$$\lambda = \frac{1}{l} \ln \frac{E_2}{E_1} \tag{3.3}$$

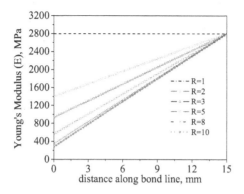

**FIGURE 3.6**   Linear variation of Young's modulus ($E$) of the adhesive layer with different modulus ratios '$R$'.

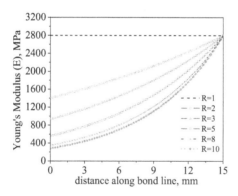

**FIGURE 3.7**   Exponential variation of Young's modulus ($E$) of the adhesive layer with different modulus ratios '$R$'.

and modulus ratio '$R$' is expressed as

$$R = \frac{E_2}{E_1} \tag{3.4}$$

where $E_1$ and $E_2$ are lower bound and upper bound Young moduli of adhesives, respectively. '$x$' is the distance measured along the bond length '$l$'. The adhesive layer used for bonding is made of FGA whose properties vary from material 1 to material 2. Material gradients are measured in terms of modulus ratio '$R$', which varies from 1 to 10. The upper bound modulus $E_2$ is taken as 2.8 GPa, and lower bound modulus $E_1$ is varied according to modulus ratio '$R$' as expressed in Eq. (3.4). The variation of elastic moduli along the bond length for linear and exponential function profiles with different elastic moduli ratios are shown in Figures 3.6 and 3.7, respectively.

The adhesive having modulus ratio 1, corresponds to mono-modulus adhesive. For modeling the tee joint with FGA, the adhesive region is assumed to have a Young's modulus changing along the $x$-axis using Eqs. (3.1) and (3.2). In FE model, the material property changes have been modeled discretely, by assigning the value of $E(x)$ at the middle for each of the elements [96] within the adhesive layer. The smooth variation of elastic moduli is ensured by keeping fine mesh (mesh size tends to zero) along the bond length.

## 3.2.1   FAILURE PREDICTION

Accurate prediction of adhesively bonded tee joint strength has been one of the challenging tasks for ensuring the structural integrity of any bonded structures. The joints joint failure mechanism includes the initial failure onset, stable failure propagation and catastrophic damage propagation. This phenomenon becomes complex and much involved when laminated FRP composite plates are used. There are large numbers of failure criteria available that have been used to predict the onset of damages in the adhesively bonded joints. Two approaches have been used for the prediction of the location of the onset of damages. One of them is based on the mechanics of materials approach, and the other follows the fracture mechanics procedures.

Failure criteria developed for isotropic materials have been utilized to predict the cohesive failure which occurs in the adhesive layer of adhesively bonded tee joint. As reported by Adams [97], the parabolic yield criterion used for the prediction of cohesive failure initiation in the tee joint is expressed below:

$$\left(\sigma_1 - \sigma_2\right)^2 + \left(\sigma_2 - \sigma_3\right)^2 + \left(\sigma_3 - \sigma_1\right)^2 + 2\left(|Y_c| - Y_T\right)\left(\sigma_{1+}\sigma_{2+}\sigma_3\right) = 2|Y_c|Y_T \quad (3.5)$$

$Y_T$ and $Y_C$ measure the yield strength in tension and compression, respectively. It may be noted that $|Y_c|$ and $Y_T$ are equal, and the above yield criterion reduces to the most familiar von-Mises cylindrical criterion which is given below:

$$\sigma_e^2 = 0.5\{\left(\sigma_1 - \sigma_2\right)^2 + \left(\sigma_2 - \sigma_3\right)^2 + \left(\sigma_3 - \sigma_1\right)^2\} \quad (3.6)$$

where $\sigma_e$ is the von-Mises stress. Coupled stress failure criterion of von-Mises as given in Eq. (3.6) is used in the present research to predict the failure onset in the adhesive layer. As expected, the adhesive layer used in the tee joint is considered to be of lower strength than the plates made of FRP composite materials.

## 3.2.2   VALIDATION OF FINITE ELEMENT MODEL

In the present FEA, the model is validated with the results available in literature by considering both the plates of the tee joint as isotropic material. The maximum principal

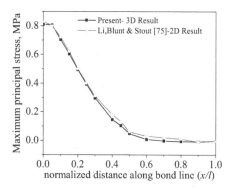

**FIGURE 3.8**   Distribution of maximum principal stresses along the adhesive layer.

stresses in the adhesive layer are evaluated and compared with those obtained from Li et al. [73]. Referring to Figure 3.8, maximum principal stress distribution within the mid-surface of the adhesive layer of the joint shows good agreement with the available results.

### 3.2.3   CONVERGENCE STUDY

Relevant error analyses and mesh refinements have been carried out to study the convergence of the result by considering the out-of-plane stress components to establish the FE model of the tee joint. A series of numerical simulations for the validated FE model with the different number of elements ($n$) have been carried out to determine the out-of-plane stress components, viz. as $\sigma_z$, $\tau_{yz}$ and $\tau_{xz}$. The out-of-plane stress values for different bond lengths ($x$) corresponding to $0.5n$, $0.75n$, $n$ and $1.2n$ are illustrated in Tables 3.4–3.6. Convergence of the full tee joint model was achieved by increasing the mesh density with the change in the out-of-plane stress values under the applied load was less than 1%. This was found to occur when the number of FEs of the model is considered as $n$.

## 3.3   DESIGN OPTIMIZATION OF TEE JOINT

Three-dimensional geometrically nonlinear FEAs are carried out for the validated model of the tee joint made of FRP composite right-angled plate. The out-of-plane stresses ($\sigma_{zz}$, $\tau_{yz}$ and $\tau_{xz}$) and von-Mises stresses ($\sigma_e$) at the midsurface of the adhesive layer of the joint are evaluated from the FEA. The von-Mises stress ($\sigma_e$) components have been calculated using Eq. (3.6). The stress values for three FRP composite materials such as Gr/E, Gl/E and B/E have different anisotropy ratios, and varied fiber orientation angles $[0]_8$, $[(0/90)_s]_2$, $[(+45/-45)_s]_2$ are determined.

### 3.3.1   LAMINATION SCHEMES AND MATERIAL ANISOTROPY

Coupled stress failure criterion of von-Mises as given in Eq. (3.5) is used to predict the failure onset in the adhesive layer. The von-Mises stress distribution on the mid-surface of the adhesive layer of tee joint with different FRP materials such as Gr/E,

TABLE 3.4

Out-of-Plane Normal Stress ($\sigma_{zz}$) Values with a Different Number of Elements ($n$)

| x | $\sigma_{zz}$ (MPa) | | | | |
|---|---|---|---|---|---|
| (mm) | 0.5$n$ | 0.75$n$ | $n$ | 1.2$n$ | % Error |
| 0 | 1.2489 | 1.2151 | 1.1753 | 1.1841 | 0.7487 |
| 1 | 0.77457 | 0.81672 | 0.82099 | 0.82911 | 0.98905 |
| 2 | 0.4993 | 0.49546 | 0.49466 | 0.49500 | 0.06873 |
| 3 | 0.2758 | 0.28922 | 0.293 | 0.29321 | 0.071672 |
| 6 | 3.80E-02 | 3.81E-02 | 3.93E-02 | 3.94E-02 | 0.2544 |
| 9 | −4.63E-03 | −4.78E-03 | −4.87E-03 | −4.88E-03 | 0.2053 |
| 12 | −1.00E-02 | −9.62E-03 | −9.50E-03 | −9.51E-03 | 0.1052 |
| 15 | −3.54E-02 | −3.80E-02 | −3.94E-02 | −3.97E-02 | 0.706014 |

TABLE 3.5

Out-of-plane Shear Stress ($\tau_{yz}$) Values with a Different Number of Elements ($n$)

| x | $\tau_{yz}$ (MPa) | | | | |
|---|---|---|---|---|---|
| (mm) | 0.5$n$ | 0.75$n$ | $n$ | 1.2$n$ | % Error |
| 0 | 0.12093 | 0.12036 | 0.1198 | 0.1199 | 0.08347 |
| 1 | 0.106918 | 0.10749 | 0.107668 | 0.107669 | 0.000929 |
| 2 | 7.56E-02 | 7.52E-02 | 7.52E-02 | 7.54E-02 | 0.265957 |
| 3 | 4.79E-02 | 4.79E-02 | 4.73E-02 | 4.73E-02 | 0.050719 |
| 6 | 8.46E-03 | 8.58E-03 | 8.69E-03 | 8.70E-03 | 0.115074 |
| 9 | 4.67E-05 | 7.42E-05 | 7.50E-05 | 7.52E-05 | 0.2.66666 |
| 12 | −1.59E-03 | −1.59E-03 | −1.61E-03 | −1.61E-03 | 0.242976 |
| 15 | −4.84E-03 | −4.91E-03 | −4.94E-03 | −4.95E-03 | 0.147696 |

TABLE 3.6

Out-of-plane Shear Stress ($\tau_{xz}$) Values with a Different Number of Elements ($n$)

| x | $\tau_{xz}$ (MPa) | | | | |
|---|---|---|---|---|---|
| (mm) | 0.5$n$ | 0.75$n$ | $n$ | 1.2$n$ | % Error |
| 0 | −6.43E-02 | −4.58E-02 | −2.84E-02 | −2.39E-02 | 0.144168 |
| 1 | −7.41E-02 | −7.42E-02 | −7.45E-02 | −7.47E-02 | 0.241546 |
| 2 | −6.06E-02 | −6.12E-02 | −6.16E-02 | −6.18E-02 | 0.272586 |
| 3 | −4.58E-02 | −4.62E-02 | −4.65E-02 | −4.65E-02 | 0 |
| 6 | −7.00E-03 | −7.11E-03 | −7.14E-02 | −7.14E-03 | 0.05042 |
| 9 | 1.05E-02 | 1.06E-02 | 1.06E-02 | 1.06E-02 | 0.321787 |
| 12 | 1.48E-02 | 1.49E-02 | 1.49E-02 | 1.50E-02 | 0.066921 |
| 15 | 1.32E-02 | 1.24E-02 | 1.17E-02 | 1.18E-02 | 0.854700 |

Gl/E, B/E and varied fiber orientation angles, viz. $[0]_8$, $[(0/90)_s]_2$, $[(+45/−45)_s]_2$, are shown in Figures 3.9–3.11.

It is seen that the tee joint with angle ply lamination scheme and the region having the highest magnitude of von-Mises stress is prone to the onset of failure in the adhesive layer. As seen from Figure 3.9a–c, tee joint made of Gr/E, the difference of magnitudes of von-Mises stress for $[0]_8$, $[(0/90)_s]_2$ is insignificant, whereas the highest stress occurs for $[(+45/−45)_s]_2$ laminate. The same situation is observed for the tee joint made of Gl/E and B/E composites. The von-Mises stress component for angle ply laminates is approximately double that of the other two laminates, irrespective of material anisotropy. It is obvious for angle ply laminates that the anisotropic elastic modulus ratio and mutual influence coefficients for an angle ply laminate may cause a high value of stress. Further, it is observed that there is not a significant variation of stress levels across the width of the joint except around the corners of right-angled plate adjacent to the adhesive layer. This indicates that the stress concentration effects are highest at the corners, and the first failure is expected to initiate from the corner of the right-angled plate for the joint made of B/E composite plates. The above discussions show that the tee joint with angle ply laminates produces high von-Mises stress levels than other laminates, irrespective of material anisotropy. It

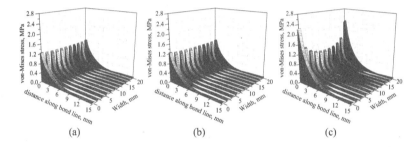

(a)  (b)  (c)

**FIGURE 3.9** Von-Mises stress distributions in the middle of the adhesive layer of the tee joint made of Gr/E Composites with varied fiber orientation angle: (a) $[0]_8$, (b) $[(0/90)_s]_2$ and (c) $[(45/−45)_s]_2$.

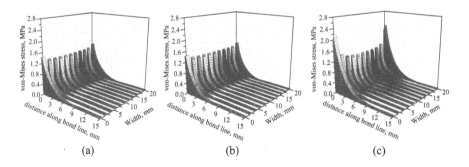

(a)  (b)  (c)

**FIGURE 3.10** Von-Mises stress distributions in the middle of the adhesive layer of the tee joint made of Gl/E Composites with varied fiber orientation angles: (a) $[0]_8$, (b) $[(0/90)_s]_2$ and (c) $[(45/−45)_s]_2$.

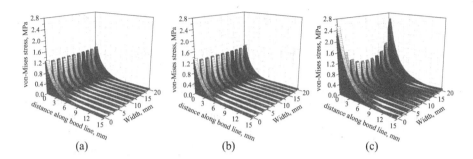

**FIGURE 3.11** Von-Mises stress distributions in the middle of the adhesive layer of the tee joint made of B/E composites with varied fiber orientation angles: (a) $[0]_8$, (b) $[(0/90)_s]_2$ and (c) $[(45/-45)_s]_2$.

is prone to failure under a given loading condition. A tee joint made of Gr/E with stacking sequence $[0]_8$ experiences lowest out-of-plane and von-Mises stress levels.

Stress analysis is carried out for tee joint (flexible base) with optimized material (Gr/E). The out-of-plane stress distribution at the midsurface of the adhesive layer for tee joint made of Gr/E composite material with stacking sequence $[0]_8$ is shown in Figure 3.12.

Like rigid-based tee joints, stresses $\sigma_z$ and $\tau_{xz}$ are not varying considerably across the joint width except around the inside corner of the right-angled plate of the tee joint with a flexible base. But 3D stress effects are more prominent across the width of the tee joint for stress level $\tau_{yz}$. The peak level of peel stresses are observed at both the free edges of the adhesive layer and remain constant over a more significant portion of the joint region. It is essential to note from Figure 3.12a that peel stress attains a maximum value at the left free edge of the adhesive layer compared with that at the right end of the joint region.

The von-Mises stress distributions on the midsurfaces of the adhesive layer of tee joint made of Gr/E with varied lamination schemes are depicted in Figure 3.13.

It is observed that the von-Mises stress magnitudes at the overlap ends are significantly high compared with the middle portion. The left free edge of the bond layer is found more vulnerable zone for the adhesive failure of the tee joint. Tee joint with flexible base experiences higher stress levels for angle ply laminate $[(+45/-45)_s]_2$ compared with unidirectional $[0]_8$, cross-ply $[(0/90)_s]_2$ composites. A comparative illustration of high-stress concentration regions based on the von-Mises stress components between the tee joints with rigid base and flexible base boundary conditions with optimized material and lamination scheme has been indicated in Figure 3.14. It is noticed that tee joint with flexible base experiences considerable stress levels.

In the view of the above-detailed discussions, it is summarized that the tee joint made of Gr/E with stacking sequence $[0]_8$ experiences the lowest out-of-plane and von-Mises stress levels and hence it should be recommended to design the tee joint structure. The design of tee joint with rigid base boundary conditions is more conservative. Hence, considering the tee joint structure with flexible base boundary conditions is more appropriate and recommended for the tee joint designer.

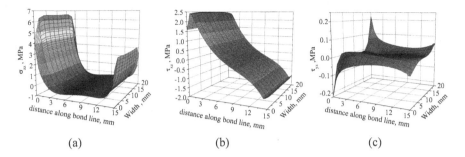

**FIGURE 3.12** Out-of-plane stress distributions at the midsurface of the adhesive layer of the tee joint (flexible base) made of Gr/E Composites with $[0]_8$ fiber orientation angles; (a) $\sigma_{zz}$, (b) $\tau_{xz}$ and (c) $\tau_{yz}$.

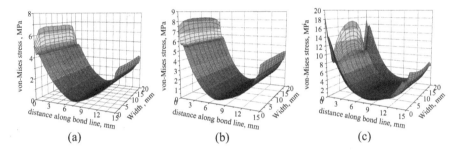

**FIGURE 3.13** Von-Mises stress distributions in the middle of the adhesive layer of the tee joint made of Gr/E Composites with varied fiber orientation angles: (a) $[0]_8$, (b) $[(0/90)_s]_2$ and (c) $[(45/-45)_s]_2$.

An attempt has been made to reduce the stress concentration further at the left free end of the adhesive layer for the tee joint structure by employing functionally graded adhesive along the bond line.

### 3.3.2 STRESS ANALYSIS OF TEE JOINT WITH FUNCTIONALLY GRADED ADHESIVE

It has been depicted experimentally and numerically that stress concentration at the end of the overlap in the joints can be reduced significantly by employing more than one adhesive along the bond line [32,34,35,98]. Their research work considers single step, discontinuous variation of elastic modulus of adhesive along the bond line. But the present research work ensures continuous smooth variation of modulus along the bond line – two material gradation profiles viz. linear and exponential function as expressed in Eqs. (3.1) and (3.2) are considered in this study.

The effect of linear and exponential material gradations with different modulus ratios on out-of-plane stress components and von-Mises stress levels induced at the middle of adhesive layer of tee joint with rigid base are shown in Figures 3.15–3.18.

It can be seen from Figure 3.15a that peel stress ($\sigma_z$) is the highest at the left free end, and its distribution along the bond length is shown for functionally graded

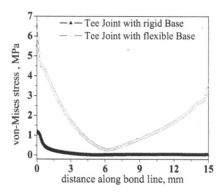

**FIGURE 3.14** A comparative illustration of von-Mises stress distributions in the middle of the adhesive layer of optimized tee joint with rigid and flexible base boundary conditions.

adhesive (linear function profile) with modulus ratios ($R$) varying from 1 to 10. Results indicate that when there is an increase in modulus ratio, the peak value of the peel stress decreases significantly by 10%–50% at the left free end of the adhesive layer by employing graded adhesive with modulus ratio $R$ varying from 2 to 5. However, 70%–100% reduction in peak stress level can be achieved by graded adhesive with a modulus ratio varying from 8 to 10 instead of mono-modulus adhesive. Thus, tee joint made with functionally graded adhesive reduces the peak value of the stresses at the overlapping end, thus reducing the possibility of failure by peel stress effect. This situation leads to an increase in strength and will delay the failure. This behavior is in qualitative agreement with numerical results [32,99,100]. The effect of linear and exponential material gradation profiles on peel stress reductions is the same except for modulus ratio 8–10. The adhesive material with a linear function profile offers more reduction in the magnitude of peak values of peel stress than stress components with the exponential profile for modulus ratio varying from 8 to 10. The effect of linear and exponential material gradations on out-of-plane shear stress reductions ($\tau_{yz}$) is more significant and similar for both profiles, as shown in Figures 3.15b and 3.16b. For modulus ratio $R=2$, peak values of shear stress level decrease by 60% compared with that of mono-modulus adhesive. Beyond this modulus ratio, peak shear stress values ($\tau_{yz}$) decrease drastically. Figures 3.15c and 3.16c show magnitudes of shear stress ($\tau_{xz}$) distribution along the bond layer. The contribution of these magnitudes toward the joint's failure is insignificant though graded adhesive tends to give rise to a small increase in peak levels.

Failure would initiate in the flexible adhesive at the ends of overlap in functionally graded adhesively bonded tee joints. A cross-section through the middle of the joint along the width has been considered to predict the onset of failure of the joint. The von-Mises stresses within the middle of the adhesive layer of the considered cross-section are evaluated and compared. The von-Mises stress distribution with mono-modulus adhesive ($R=1$) and functionally graded adhesive with linear and exponential material gradation profiles ($R=2$–10) are depicted in Figures 3.17 and 3.18. It is indicated from Figures 3.17 and 3.18 that von-Mises stress magnitude is highest at the overlapping end and reduces by 10%–50% in the graded bond line

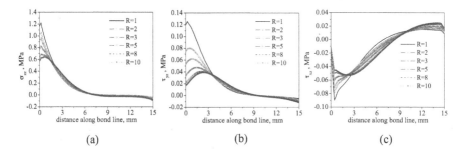

**FIGURE 3.15** Effect of functionally graded adhesive (linear function profile) on out-of-plane stresses (a) $\sigma_z$, (b) $\tau_{yz}$ and (c) $\tau_{xz}$ with varied modulus ratios '$R$'.

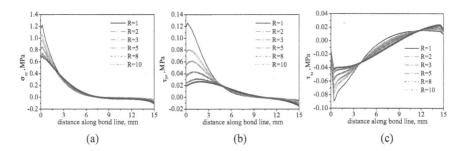

**FIGURE 3.16** Effect of functionally graded adhesive (exponential function profile) on out-of-plane stresses (a) $\sigma_z$, (b) $\tau_{yz}$ and (c) $\tau_{xz}$ with varied modulus ratios '$R$'.

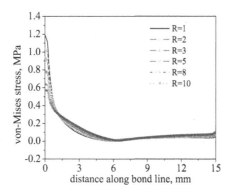

**FIGURE 3.17** Effect of functionally graded adhesive (linear function profile) on von-Mises stresses with varied modulus ratios '$R$'.

when the modulus ratio varies from 2 to 5. However, 80%–100% reduction in the magnitude of von-Mises stresses can be achieved by using graded bond lines with varying modulus ratios from 8 to 10. Similar behaviors are observed by Kumar and Pandey [98] and Temiz [100].

Similarly, the effects of optimized material gradation function profile (linear) on out-of-plane stresses induced in the tee joint with a flexible base are shown in

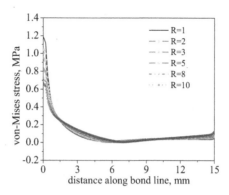

**FIGURE 3.18**   Effect of functionally graded adhesive (exponential function profile) on von-Mises with varied modulus ratios '*R*'.

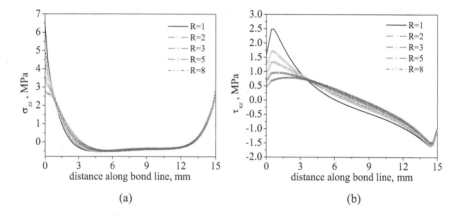

**FIGURE 3.19**   Effect of functionally graded adhesive (linear function profile) on out-of-plane stresses (a) $\sigma_z$ and (b) $\tau_{xz}$ with varied modulus ratios '*R*'.

Figure 3.19. Excellent reductions in peak values of out-of-plane stresses are noticed inflexible based tee joint.

Overall, the above discussion indicates that functionally graded adhesive along bond length improves joint strength and performance significantly and failure onset in adhesive layer can be delayed drastically.

## 3.4   SUMMARY

The present study shows a significant effect of material anisotropy and fiber orientation angle on the distribution of out-of-plane and von-Mises stress components in the tee joint. It is also observed that maximum values of out-of-plane stress component in the tee joint can be reduced by using functionally graded adhesive along the bond line. This is a desirable recommendation for the designers while designing a tee joint. The following specific conclusions made out of the present research are given below:

- A tee joint with angle ply composite plates produces high out-of-plane and von-Mises stress levels compared with other lamination schemes and is prone to failure and hence should be avoided by the designers.
- A tee joint made of Gl/E experiences highest von-Mises stress level for stacking sequences $[0]_8$ and $[(0/90)_s]_2$ compared with other composites and are undesirable for tee joint structure.
- A tee joint made of Gr/E with stacking sequence $[0]_8$ has experienced the lowest out-of-plane and von-Mises stress levels and, hence it should be recommended for the design of tee joint structure.
- The design of tee joint with rigid base boundary conditions is more conservative. Hence, considering the tee joint structure with flexible base boundary conditions is more appropriate and recommended for the tee joint designer.
- Tee joint made with functionally graded adhesive reduces the peak value of the stresses at the overlapping end, thus reducing the possibility of failure by peel stress effect. This situation leads to an increase in strength and will delay the failure. Hence, the designer recommends a functionally graded adhesively bonded tee joint due to its improved strength.
- The adhesive material with a linear function profile reduces the magnitude of peak values of peel stress than that of stress components with an exponential profile. Hence, a bond layer with a linear gradation profile is recommended for joint design.

# 4 Study of Damage Growth in Functionally Graded Adhesively Bonded Double Supported Tee Joint

## 4.1 INTRODUCTION

Adhesive bonding for out-of-plane joints of fiber-reinforced polymeric (FRP) composites has been one of the most important and evolving technologies for many structural applications in a great variety of industries. Adhesive bonding can offer improved performance and substantial economic advantages compared with other joining methods such as mechanical fastening, welding, brazing and soldering. The ability to join dissimilar materials (such as laminated composites), joining of thin sheets and joining of materials with complex geometrical configurations has made the adhesive bonding more attractive over the other methods of joining. A smoother load transfer between the connecting members helps in lowering the localized stress concentrations compared with mechanical fasteners. It offers the potential of reduced weight and cost, and it has found widespread acceptance in many engineering fields.

Tee joint configuration represents one of the most critical joining schemes, commonly adopted in aeronautical and marine fields, to efficiently assemble composite laminates in primary structural elements [25,101]. Such type of joint undergoes different failure modes as shown in Figure 4.1a, which are listed below:

- Adhesion or interfacial failure at the base/support plate–adhesive interfaces caused by excessive peel and shear stresses at critical locations
- Cohesive failure within the adhesive layer
- Out-of-plane base/support plate failure in laminated FRP composite plates caused by interlaminar stresses

The present research focuses on bond line failure, including interfacial and cohesive failure. There has been practically minimal effort to analyze the interfacial failure, known as adhesion failure or delamination/debonding damages caused by interfacial peel and shear stresses. Such damages occur at the bi-material interfaces, that is, at the base/support plate interface and adhesive in case of bonded tee joints.

DOI: 10.1201/9781003201113-4

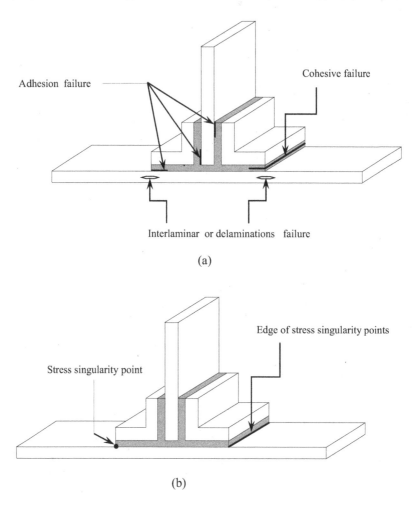

**FIGURE 4.1**   (a) Failure modes in adhesively bonded tee joint of laminated fiber-reinforced polymeric composites, and (b) Initiation of interfacial failures from stress singularity points.

Interfacial failure occurs on a macro-scale when surface preparation or material qualities are poor, and, consequently, this mode of failure cannot be ignored. Although joints are assumed to be manufactured to specifications, adhesion failures are expected to initiate from the edges of stress singularity points due to the aforementioned reasons. They would propagate along with the base/support plate–adhesive interfaces, as shown in Figure 4.1b.

Davies et al. [102] developed a new interface element with a monotonic (exponential) force/displacement to model the onset of delamination/debonding in composite tee joint structures and subsequent propagation. It was observed that the interface element initiates and propagates correctly without needing an interfacial flaw.

Dharmawan et al. [103] studied the structural integrity and damage tolerance of typical marine composite tee joints made of glass FRP. They carried out finite element analysis (FEA) to investigate the effects of disbonds between the filler and over laminate on the strain distribution in the joint. The same authors noticed that particular defects led to large changes in the strains in the tee joint, which would encourage disbond progression.

Li et al. [24] conducted FEA for glass FRP tee joint commonly used in composite marine vessels. The virtual crack closure technique was used to investigate the fracture behavior of tee joint structures with pre-embedded disbond with varied sizes at different locations under a straight pull-off load.

The same authors evaluated strain energy release rate (SERR) at disbond tips to assess fracture behavior and damage criticality of such structures.

Kesavan et al. [56,104] presented a study of composite tee joint strain distribution under tensile pull-off loads and determined the presence and extent of disbonds. FEA was conducted by embedding delaminations of varied sizes at various locations in the structure. The results were validated experimentally, and the resulting strain distribution from the numerical simulation was pre-processed using a damage relativity assessment technique. The same authors used an artificial neural network to determine the extent of damage.

Baldi et al. [105] performed experiments to study the flexural behavior of composite tee joints in the presence of a structural adhesive film between the joined sub-laminates. A numerical approach based on the cohesive zone model was applied to simulate the onset and subsequent propagation of interlaminar damages in adhesive and interlaminar layers.

It is essential to retard damage growth in the joint structure to have enhanced structural integrity of the out-of-plane joint structure. More recently, there have been some studies in the improvement of the joint strength by making the use of the mixed adhesive joint (with ductile and brittle adhesives) [32–35,98,100,106]. The mixed adhesive joint technique can be considered a rough version of a functionally graded material.

Sancaktar and Kumar [86] used rubber particles to modify the adhesive locally at the ends of the overlap, and Stapleton et al. [7] used glass beads strategically placed within the adhesive layer to obtain different densities and change the stiffness along the overlap. Carbas et al. [89] performed FEA for a single-lap joint with varied adhesive stiffness along the overlap, being maximum in the middle and minimum at the ends of overlap. Results showed higher strength for functionally graded joints than that of joints with mono-modulus adhesive. Nimje and Panigrahi [107] made an effort to improve the strength of the out-of-plane joint by using functionally graded adhesive along the bond line. Two material gradation function profiles (linear and exponential) were used to grade the bond line. The same researchers have performed the parametric study with varied modulus ratios to show their influence on out-of-plane and von-Mises stresses along the bond length. To date, 2D and 3D stress analyses have been done by using functionally graded adhesive along the bond line to reduce peel and shear stress concentrations at the ends of overlap by which the strength of the joint improved significantly.

From the above-said analysis and discussion, the obvious fact is that the adhesively bonded tee joints of laminated FRP composite plate are prone to defects/damages due to various factors such as aging, corrosion, delamination between the plies, cohesive failure within the adhesive layer or interfacial failure between the adhesive layer and base/support plate. Usually speaking, these damages/defects reduce the strength, stiffness of the joint and load-carrying ability of the structure and alter the structure's response to external loads. Therefore, the present research focuses on the damage propagation behavior of functionally graded adhesively bonded double supported tee joints made of laminated FRP composite plates. 3D FE simulations have been performed to evaluate peel and shear stress. Subsequently, a coupled stress failure criterion has been used to determine the failure indices to predict damage onset location. Damage propagation analyses have been studied using the SERR values along damage fronts based on fracture mechanics principles. Furthermore, an attempt has been made to improve damage growth resistance by using functionally graded adhesive with an appropriate function profile along the bond line. Several numerical simulations have been carried out to indicate the effect of various modulus ratios on damage propagation rate for varied damage lengths.

## 4.2 NUMERICAL SIMULATION OF DOUBLE SUPPORTED TEE JOINT USING FINITE ELEMENT ANALYSIS

Numerical simulations can play an essential role to assist the design of damage-tolerant joints. They can reduce the number of experiments needed to evaluate the structural response in the actual working environment and define optimal design solutions. In such a context, 3D nonlinear FEAs have been carried out to assess the structural behavior of adhesively bonded double supported tee joint of laminated FRP composites having embedded interfacial failures.

The adhesively bonded tee joint is shown in Figure 4.2, which consists of a vertical plate, a base plate and two double supports, in which the vertical plate is bonded

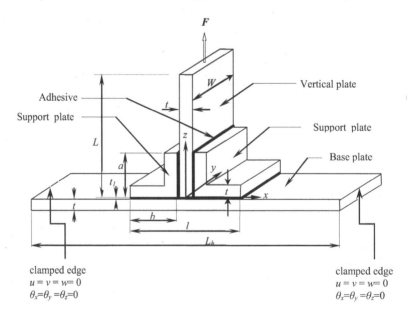

**FIGURE 4.2**   Configuration of double supported tee joint.

to the base plate with the help of two corner supports. Dimensions of the considered tee joint are given in Table 4.1.

The vertical plate, support plates and base plate consist of $[0]_8$, graphite/epoxy FRP composite laminates. The thickness of each ply is taken as 0.25 mm. The layer-wise orthotropic properties of FRP composite laminates are shown in Table 3.2, and strength properties are shown in Table 4.2. The joint is subjected to an out-of-plane load of '$F$' equaling to 50 N in $z$-direction through the vertical plate. Two types of adhesives, namely mono-modulus and functionally graded adhesives, are used to bond the vertical plate with two support plates and a base plate. The material properties of the mono-modulus adhesive are given in Table 4.3. The details of functionally graded adhesive considered in the present research are extensively explained in a proceeding section.

Two bond layers, namely horizontal and vertical, are used to bond the vertical plate with the base plate of the joint. Both bond layers are graded with functionally graded adhesive. The justifying facts for the use of linear material gradation function profile are elaborately discussed in Section 3.3.2. Hence, the FGA is implemented through continuous variation of elastic modulus along the bond line, which is governed by a linear function profile using Eq. (3.1).

### Table 4.1
### Dimensions of Tee Joint Model [73, 109]

| Parameters | Dimensions (mm) |
| --- | --- |
| Vertical support length, $a$ | 15 |
| Horizontal support length, $b$ | 15 |
| Bond length, $l$ | 33 |
| Plate thickness, $t$ | 2 |
| Adhesive thickness, $t_1$ | 0.5 |
| Vertical plate length, $L$ | 40 |
| Base plate length, $L_h$ | 100 |
| Joint width, $W$ | 20 |

### Table 4.2
### Strength Properties of Gr/E (T300/934) Composite Plates [27, 29]

| | |
| --- | --- |
| $Z$ (Out-of-plane normal (transverse) strength) | 49 (MPa) |
| $S$ (Out-of-plane shear strength) | 2.55 (MPa) |

### Table 4.3
### Elastic and Strength Properties of Epoxy Adhesive [27, 29]

| $E$ | 2.8 (GPa) | Strengths | $Y_T$ | 65 (MPa) |
| --- | --- | --- | --- | --- |
| $\upsilon$ | 0.4 | | $Y_C$ | 84.5 (MPa) |

Because of symmetry of the double supported tee joint, material gradation properties of the horizontal bond line are shown for half portion of joint, which is shown in Figure 4.3. Eq. (3.1) can be used for material gradation profile for the vertical bond line by replacing '*x*' with '*z*' and '*l*' by '*a*', where '*z*' is the distance measured along the vertical bond line and '*a*' denotes the bond line length.

Gradation properties along the vertical bond line are depicted in Figure 4.4. Based on the previous study [107] and from stress distributions, flexible adhesive having low values of elastic modulus is used in the tee joint at the center and overlap ends of support plates, whereas stiffest adhesive having the highest values of elastic modulus is used at the center of the horizontal support plate. Stiffer adhesive following linear profile, as shown in Figure 4.3 has been used in between. The flexible adhesive is used near the base plate for the vertical bond layer, and stiffer adhesive is used at the vertical overlap end. In the FE model, the changes of material property have been modeled discretely by assigning each element the value of *E(x)* at the middle of each element.

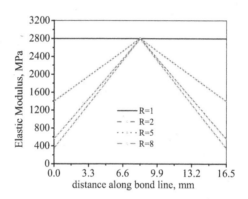

**FIGURE 4.3**   Gradation of elastic modulus along the horizontal bond line for different modulus ratio '*R*'.

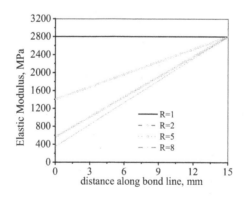

**FIGURE 4.4**   Gradation of elastic modulus along the vertical bond line for different modulus ratio '*R*'.

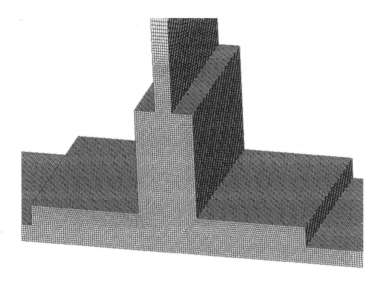

**FIGURE 4.5**   Zoomed view of finite element model for double supported tee joint.

Fine mesh is adopted to model the adhesive layer to achieve continuous and smooth variation of elastic modulus. Error and convergence studies have been done extensively and shown in Section 3.2.3. The same meshing scheme has been adopted to model double supported tee joint.

Considering these points, volume elements designated as SOLID 45 and SOLID 46 of ANSYS are used for modeling the adhesive layer and FRP composite plates, respectively. One element is used to model the composite plate through the thickness, whereas two elements are used to model the adhesive layer. These are 3D isoparametric solid elements. SOLID 46 is a layer version of the eight-node structural solid (SOLID 45) element, designed to model the laminated FRP composite plates ply-by-ply. Orthotropic material properties have been considered for each ply. These elements have eight nodes, and each node possesses three translational degrees of freedom. The FE meshing scheme used for modeling and simulating damages in the considered tee joint structure is shown in Figure 4.5.

## 4.3   INTERFACIAL FAILURE PROPAGATION ANALYSIS IN TEE JOINT

3D FEAs with geometric nonlinearity have been carried out for tee joint to predict the location of initiation of damages at the interfacial surfaces and within the adhesive layer. In addition, damage propagation analyses are performed for functionally graded adhesively bonded tee joint structures. Out of two boundary conditions, rigid base and flexible base, the design of the tee joint with rigid base boundary condition is more conservative [75,107]. The practical applications of tee joints for stiffening wing-skin of aircraft structures and hull-bulkhead joint in ship structure flexible boundary conditions are more appropriate and, therefore, considered in the present

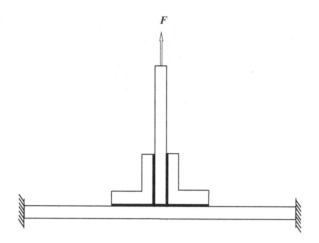

**FIGURE 4.6**  Loading and boundary conditions used for numerical simulation.

research work. The tee joint with flexible base boundary condition is depicted in Figure 4.6. Accordingly, both ends of the base plate are fixed inflexible boundary conditions. A total out-of-plane load 'F' equal to 50N has been applied to the tee joint structure in 100 equal sub-steps to achieve geometrical nonlinearity.

Adhesively bonded tee joint experiences two important modes of mechanical failure: (i) interfacial failure also known as adhesion failure, which occurs at the interface of adhesive and support/base plate and (ii) cohesive failure within the adhesive layer apart from the failure or damage due to delamination in composite plates. This research is concerned with a detailed understanding of the mechanics of interfacial failure in out-of-plane joints. Tsai and Wu [60] have predicted the onset of failures over the bond line interfacial surfaces using coupled failure criterion through computation of failure indices 'e'. Accordingly, a failure surface in the stress space can be represented in the following form:

$$f(\sigma_k) = F_i\sigma_i + F_{ij}\sigma_i\sigma_j = 1 \tag{4.1}$$

where $i, j, k = 1, 2, ..., 6$, and $F_i$ and $F_{ij}$ are strength tensors.

In the tee joint structure, the interlaminar or out-of-plane stresses are responsible for the initiation of interfacial failures. Hence, these stress components only have been used in Eq. (4.1) to determine the failure index values 'e' and are given by

i. Interfacial failure in tension, for $\sigma_z > 0$;

$$\left(\frac{\sigma_z}{Z_T}\right)^2 + \left(\frac{\tau_{xz}}{S_{xz}}\right)^2 + \left(\frac{\tau_{yz}}{S_{yz}}\right)^2 = e^2 \begin{cases} e \geq 1, & \text{failure} \\ e < 1, & \text{no failure} \end{cases} \tag{4.2}$$

ii. Interfacial failure in compression, for $\sigma_z < 0$;

$$\left(\frac{\sigma_z}{Z_C}\right)^2 + \left(\frac{\tau_{xz}}{S_{xz}}\right)^2 + \left(\frac{\tau_{yz}}{S_{yz}}\right)^2 = e^2 \begin{cases} e \geq 1, & \text{failure} \\ e < 1, & \text{no failure} \end{cases} \tag{4.3}$$

Similarly, the failure index of the adhesive layer is formulated by cohesive failure philosophy. As reported by Adam [97], parabolic yield criterion is expressed as

$$(\sigma_1 - \sigma_2)^2 + (\sigma_2 - \sigma_3)^2 + (\sigma_3 - \sigma_1)^2 + 2(|Y_C| - Y_T)(\sigma_1 + \sigma_2 + \sigma_3) = 2e|Y_C|Y_T \quad (4.4)$$

Using Eqs. (4.2)–(4.4), failure indices are evaluated for three surfaces viz. two interfacial surfaces and midsurface of the adhesive layer, respectively. $Y_T$ and $Y_C$ measure the yield strength in tension and compression, respectively. Based on the magnitudes of '$e$', the critical locations for the onset of damage in the tee joint structure have been identified. Subsequently, the damage propagation behavior of tee joint has been studied using SERRs ($G_I$, $G_{II}$, $G_{III}$).

Due to interfacial failure emanating from the critical locations, the damage propagation behavior has been modeled. Their propagations are governed by the individual modes of SERR along the interfacial failure front for the considered laminated FRP composite tee joint. In the laminated FRP composite plates of tee joint, due to their inherent complexities due to geometrical, loading and material properties, exact closed-form expressions for SERR are not possible. The singularity of crack-tip stress field in an orthotropic media is quite different from that of the conventional square-root singularity at the crack-tip inhomogeneous isotropic material system. This leads to evaluating interlaminar fracture energy released due to the propagation of the existing interfacial failure by a very small amount [108]. The SERR procedure is suitable for assessing damage propagation behavior because it is based on a sound energy balance principle implying its robustness. In addition, mode separation of SERR is possible. Irwin's crack closure theory has been followed to evaluate individual modes of SERRs. This aspect is very important as in most of cases, and the fracture mechanism is a mixed-mode phenomenon in multidirectional laminated FRP composites.

Based on the magnitudes of failure indices, interfacial failure is expected to trigger at the interface of the base plate and adhesive layer from both ends of the horizontal bond line as shown in Figure 4.7. Accordingly, the configuration for evaluation of modified crack closure integral (MCCI) applied to adhesively bonded tee joint for computation of SERR along the embedded interfacial failure length '$a$' existing at the base plate and adhesive layer interface is shown in Figure 4.8.

Except over the interfacial failure region, multipoint constraints are imposed on the nodes along the interfacial failure front. Furthermore, it has been assumed that the interfacial failure plane is the weakest, and interfacial failure will propagate parallel to the $x$–$y$ plane. Thus, the possibility of out-of-plane propagation is ignored. The three components of SERR, viz. $G_I$, $G_{II}$ and $G_{III}$, have been evaluated using MCCI and have been used as parameters for assessing the damage propagation characteristics [108]. The central point of focus of any SERR analysis is the evaluation of interlaminar stresses $\sigma_z$, $\tau_{xz}$ and $\tau_{yz}$ along the interfacial failure front and the displacement fields around it. Then the strain energy release rates along the interfacial failure front can be calculated from those stresses and displacement fields using Irwin's theory of crack closure. The strain energy released by the propagation of interfacial failure of length $a$ to $a + \Delta a$ is given by

$$W = \frac{1}{2} \int_{a}^{a+\Delta a} \int_{-\Delta a/2}^{\Delta a/2} \sigma(x,y) \times \delta(x - \Delta a, y) \, dx \, dy \quad (4.5)$$

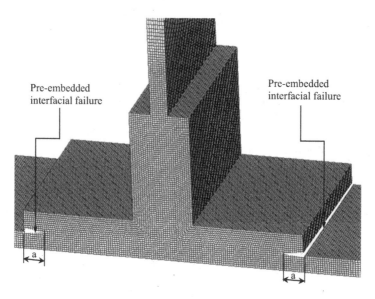

**FIGURE 4.7** Pre-embedded interfacial failure existing between the interface of the base plate and the adhesive layer (initiates from the free edges) of the tee joint.

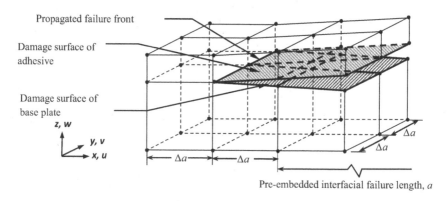

**FIGURE 4.8** Modified crack closure integral applied to the tee joint for computation of strain energy release rate along the interfacial failure front.

where $\delta(x - \Delta a, y)$ is the crack opening displacement between the top and bottom interfacial failure surfaces and $\sigma(x, y)$ is the stress at any point on the interfacial failure front required to close the delaminated area. Then, the strain energy release rate ($G$) is obtained as

$$G = \lim_{\Delta a \to 0} \frac{W}{\Delta A} \qquad (4.6)$$

where $\Delta A$ represents the interfacial failure propagated area and equals to one element area in $x$–$y$ plane, that is, $\Delta a \times \Delta a$ for the present case. The MCCI has the advantage of mode separation of strain energy release rate, which will help for a qualitative analysis of damage propagation behavior. Accordingly, the three components of strain energy release rates $G_I$, $G_{II}$ and $G_{III}$ for Modes $I$, $II$ and $III$ can be expressed as follows:

$$G_I = \lim_{\Delta a \to 0} \frac{1}{2\Delta A} \int_{a}^{a+\Delta a} \int_{-\Delta a/2}^{\Delta a/2} \sigma_z(x,y) \times [w_T(x - \Delta a, y) - w_B(x - \Delta a, y)] \, dx \, dy \quad (4.7)$$

$$G_{II} = \lim_{\Delta a \to 0} \frac{1}{2\Delta A} \int_{a}^{a+\Delta a} \int_{-\Delta a/2}^{\Delta a/2} \tau_{xz}(x,y) \times [u_T(x - \Delta a, y) - u_B(x - \Delta a, y)] \, dx \, dy \quad (4.8)$$

$$G_{III} = \lim_{\Delta a \to 0} \frac{1}{2\Delta A} \int_{a}^{a+\Delta a} \int_{-\Delta a/2}^{\Delta a/2} \tau_{yz}(x,y) \times [v_T(x - \Delta a, y) - v_B(x - \Delta a, y)] \, dx \, dy \quad (4.9)$$

The 3D nature of the stress field on different interfacial surfaces, that is, (i) support plates and adhesive layer, (ii) middle of the adhesive layer and (iii) base plate and the adhesive layer of the joint have been used for finding out the region for initiation of failures in the joint. The failure indices 'e' have been computed using Tsai-Wu's coupled stress failure criterion given in (Eqs. 4.2 and 4.3). The variations of 'e' are shown in Figure 4.9. It is observed that 'e' values are the highest at stress singularity points, that is, at the interface of the base plate and adhesive layer.

Referring to Figure 4.9, it is clearly observed that the interfacial failure onset triggers at the interface of the base plate and adhesive layer from both ends of horizontal bond length. Apalak et al. [109] presented a stress-based double supported tee joint design. They found stress concentrations around the free ends of horizontal plate–adhesive interface for similar geometry and tensile loading condition. Da Silva and Adams [110] found the locus of cracks in failed double supported tee joint under pull-off loading is situated in the same region. Thus, failure initiation location predicted in the present research work is in qualitative agreement with experimental and numerical evidence available in published literature.

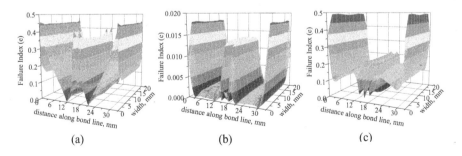

**FIGURE 4.9** Failure index 'e' distribution in laminated fiber-reinforced polymeric composite tee joint on different surfaces between (a) support plate and adhesive layer, (b) adhesive layer and (c) base plate and adhesive layer.

The failure analyses indicate the critical locations for the onset of interfacial failure. The joint vulnerability can be ascertained by pre-embedding the damages at the critical locations of the tee joint (Figure 4.8) and allowing it to propagate. Individual modes of SERR, $G_I$, $G_{II}$ and $G_{III}$ considered as fracture parameters governing the propagation of damages, have been computed using Eqs. (4.7)–(4.9) along the interfacial failure fronts.

Figure 4.10a–c exhibits the variations of $G_I$, $G_{II}$ and $G_{III}$ with varied interfacial failure lengths '$a$'. Referring to Figure 4.10a–c, it is observed that the interfacial failure front will propagate at an almost constant rate except at the edges. Near the free edges $G_I$, $G_{II}$ values are significantly less than those in the center of the interfacial failure front. This behavior is in qualitative agreement with Dharmawan et al. [111]. From the boundary region, mode *III* SERR ($G_{III}$) values are significantly lower. But at the middle portion, interfacial failure propagation is not due to individual modes. However, it will be propagated under mixed-mode SERR values go on increasing, which indicates that structural integrity of the tee joint will go on reducing as the interfacial failure propagates.

On comparing the individual modes of SERR, $G_{III}$ is insignificant except at free edges. Initially, $G_I$ and $G_{II}$ are the same when interfacial failure starts propagating. For increased damage length, $G_{II}$ plays a leading role in propagating interfacial failure. The contributions of $G_I$ are significant for all values of interfacial failure length. Hence,

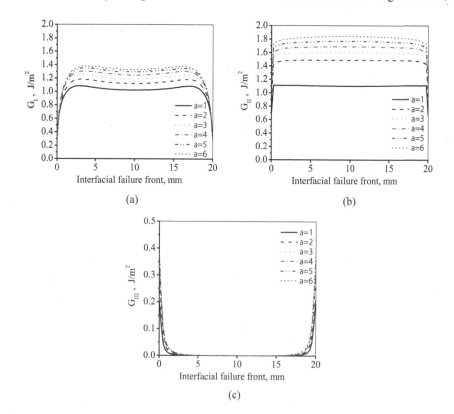

**FIGURE 4.10** Variations of individual modes of strain energy release rate along the interfacial failure front for varied pre-embedded interfacial failure lengths '$a$' at the interface of the base plate and adhesive: (a) $G_I$, (b) $G_{II}$ and (c) $G_{III}$.

it can be said that interfacial failure propagates under mixed-mode conditions. Therefore, total SERR $(G_T)$ is considered as governing parameter for interfacial failure propagation.

The desirable intention of the tee joint designer is to retard the interfacial failure propagation rate to enhance the structural integrity of the tee joint. As a result, the strength and lifetime of the tee joint structure can be significantly improved. The present research made efforts to retard interfacial failure propagation rate by using functionally graded adhesive along the bond line. The effects of the graded bond line on propagation rate are extensively discussed in the next section.

## 4.4   EFFECT OF MATERIAL GRADATION OF ADHESIVE ON DAMAGE GROWTH

In the present work, the adhesive modulus is continuously and smoothly varied along the bond line using the appropriate linear function profile (Eq. 3.1). As reported earlier in Section 4.3 and from numerical and experimental evidences [109,110], it is expected that failure initiation takes place at the interface of the base plate and adhesive layer from both ends of horizontal bond length. Figure 4.11a–d exhibit the total SERR $(G_T)$ variations with modulus ratios $R = 1, 2, 5, 8$ for varied interfacial

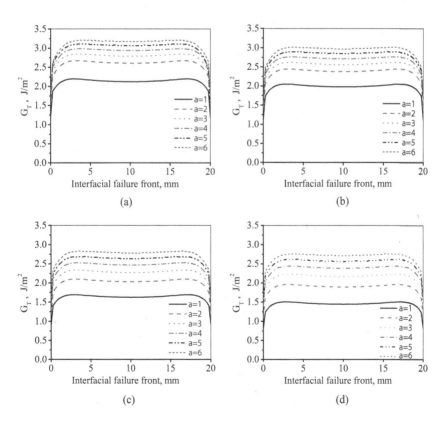

**FIGURE 4.11**  Total strain energy release rate along the interfacial failure front for varied pre-embedded interfacial failure lengths '$a$' in functionally graded adhesively boned tee joint with various modulus ratios $R$: (a) $R = 1$, (b) $R = 2$, (c) $R = 5$ and (d) $R = 8$.

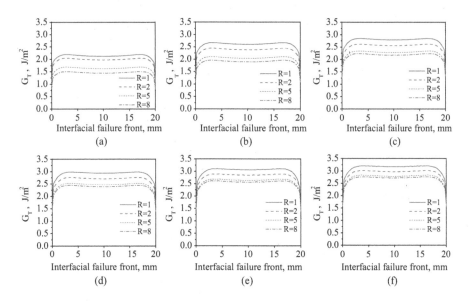

**FIGURE 4.12** Effect of functionally graded adhesive on strain energy release rate at various pre-embedded interfacial failure lengths: (a) $a = 1$ mm, (b) $a = 2$ mm, (c) $a = 3$ mm, (d) $a = 4$ mm, (e) $a = 5$ mm and (f) $a = 6$ mm.

failure lengths $a = 1$–6 mm. It is observed that the interfacial failure propagation rate ($G_T$) is curving toward the free edges of the failure front. This behavior with graded adhesive is found similar to mono-modulus adhesive ($R = 1$).

Referring to Figure 4.11a–d, it may be noted that the loci of total SERR ($G_T$) values are continuously reducing for varied gradation material properties of the adhesive irrespective of different failure length. This indicates that the damage driving forces are continuously decreasing with respect to modulus ratios. Figure 4.12a–f shows the effect of functionally graded adhesive with various modulus ratios on SERR along the failure front of any interfacial failure lengths.

When there is an increase in the modulus in the direction of crack front propagation, SERR values at the base plate interface and graded adhesive are lower relative to the SERR of the crack front at the interface of base plate mono-modulus adhesive ($R = 1$). Reductions in damage driving forces are noticed for all interfacial failure lengths as shown in Figure 4.12a–f. This behavior is in qualitative agreement with the results observed by Ravi Chandran and Barsoum [93] for crack growth analysis of graded plates.

Exact reduction in the magnitudes of SERR at the middle of interfacial failure front due to graded adhesive with varied modulus ratios is clearly indicated in Figure 4.13.

Here total SERR ($G_T$) is normalized by $G_{T, homo}$. $G_{T, homo}$ stands for total $G$ with homogeneous adhesive material. This normalized value is plotted against the material nonhomogeneity parameter, that is, modulus ratio ($R$). $G_T$ value at the middle of interfacial failure front with nonhomogeneous graded adhesive material is smaller than that of the homogeneous adhesive material.

As the material nonhomogeneity increases, the difference between $G_T$ and $G_{T, homo}$ increases. For a specific value of modulus ratio $R$, the effect of material gradation

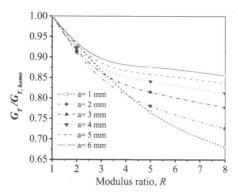

**FIGURE 4.13**  Effect of adhesive gradation on strain energy release rate at the mid of pre-embedded interfacial failure front in tee joint.

of adhesive on $G_T$ is more intense for shorter interfacial failure lengths. This linear material gradient profile indicates excellent SERR reduction for shorter interfacial failure lengths, which are necessary design characteristics of functionally graded adhesive to arrest the damage growth. This behavior of graded structure is in qualitative agreement with many researchers [112] who showed that material property gradation could cause a decrease in the amplitude of mode-I stress intensity factors under both mechanical and thermal loading. The other researcher [113] applied strain energy density criterion to crack problem in graded materials. The effect of property gradation on critical conditions for onset of crack growth is discussed and shown that graded material can offer more resistance to crack growth and suppress crack growth in some situations. Overall, the above discussion indicates that the potential use of functionally graded adhesive can be explored widely for improved damage growth resistance in adhesively bonded tee joint structure.

## 4.5 SUMMARY

Due to interfacial failure analyses of laminated FRP composite tee joint structure, critical locations for damage onset have been identified. The behavior of tee joint with pre-embedded damages has been studied by three-dimensional FEAs with varied interfacial failure lengths. MCCI has been used to evaluate damage growth driving forces (SERR). The detailed FEA indicates that damage initiation takes place at the interface of the base plate and adhesive layer from both ends of the horizontal bond line of the tee joint subjected to pull-off load. Furthermore, damage propagates under mixed-mode condition. The variation of total SERR across the damage front is constant except at the free edge. Thus, the nonuniform SERR distribution leads to conclude anisotropic and non-self-similar damage propagation characteristics. The rate of damage propagation in the central portion of the tee joint is at a higher rate than that in the free edge. Therefore, straight damage front may grow into a curved front. Furthermore, the effect of functionally graded bond line in tee joint structure shows a significant reduction in SERR. The characteristic feature of high damage growth resistance of graded tee joint structure improves the service life and can be recommended for joint designers.

# 5 Functionally Graded Adhesively Bonded Joint Assembly under Varied Loading

## 5.1 INTRODUCTION

Tee joints made of fiber-reinforced polymeric (FRP) composites are finding a wide range of applications in the aerospace industries because these materials possess high specific strength and specific stiffness. Tee joints are abundantly used in wing skins, spars, ribs, fuselage bulkheads, longerons, etc. Bonded/fastened joints are required for composite assemblies to transfer load efficiently. The advantages of adhesively bonded composite joints include more uniform stress distribution, better fatigue properties and reduced structural weight. Therefore, tee sections having adhesively bonded joints play a significant role in aerospace, wind turbine and ship designs, etc. The assembly of complex composite structures in aerospace industries dictates efficient out-of-plane joining methods such as tee joint structures. Tee joint structures are extensively used in fuselage bulkhead-to-skin, rib-to-skin, spar-to-skin and longeron-to-skin interfaces. T-stiffeners are also widely used in aircraft wings to prevent buckling during wing loading. The behavior and performance of the tee joint structure must be known to ensure joint reliability and structural integrity.

Shenoi and other researchers [72,114] have investigated adhesively bonded composite tee joints (hull-bulkhead structures) in small boats using a design tool based on experimental and numerical analysis. The same authors have studied the effect of joint geometry on an out-of-plane load-carrying capacity of joints.

Apalak and colleagues [77,109,115,116] carried out stress and stiffness analysis of double supported tee joint made of steel plates under different loading and boundary conditions. The same authors investigated the effects of various geometry parameters on the peak adhesive stresses and joint stiffness and recommended appropriate support dimensions relative to plate thickness. They also showed that geometrically nonlinear analysis had a considerable effect on the deformation and stress states of both adherends and adhesive layers. Their investigations were carried out using the incremental finite element (FE) method and are limited to isotropic metallic adherends and mono-modulus adhesive.

Da Silva and Adams [110] measured the strength of tee joints made of steel with aluminum-filled structural epoxy adhesive. They observed significant plastic

DOI: 10.1201/9781003201113-5

deformation of steel plates before the failure of adhesive. Theotokoglou et al. [117,118] have performed an experimental and numerical study of composite sandwich tee joints subjected to pull-of-load. They have found the presence of both geometric and material nonlinearities and geometric nonlinearities were caused by out-of-plane loading.

The requirement for reduced structural weight and ever-increasing demand for more efficient aerospace structures has driven the development of adhesively bonded tee joints. However, adhesively bonded tee joints are sensitive to peel and through-the-thickness stresses. Currently, there are intense investigations in joint strength improvements. This strength improvement has been obtained through different types of approaches such as altering adherend geometry [81,82], adhesive geometry [83], spew geometry [80,85] and ply dropping technique. Tapering adherend near an overlapping end is simple and effective in reducing the peak peel stress and peak shear stress. For metal adherends, tapering can be achieved via the machining process, while for a composite adherend, tapering can be realized by dropping plies gradually near the overlapping end. Although end tapering is simple and effective, it can be labor intensive and less cost-effective, especially for composite adherend [29]. Another technique to improve the joint strength is using more than one adhesive (so-called mixed adhesive joints). This approach consists of using stiff and robust adhesive in the middle of the joint and a flexible and ductile adhesive near the free edges of the bond layer to relieve the high-stress concentrations at the free ends of the joint region [32–35,98,100,106]. These researchers have shown that the mixed adhesive technique gives joint strength improvements with a brittle adhesive alone in all cases. If the ductile adhesive has a joint strength lower than that of the brittle adhesive, a mixed adhesive joint with both adhesives gives a joint strength higher than the joint strength of the adhesives used individually. The mixed adhesive joint technique can be considered a rough version of a functionally graded material. The ideal would be to have an adhesive functionally modified with properties that vary gradually along the overlap allowing an accurate uniform stress distribution along the bond layer of the joint. More recently, few investigators [7,46,99,119,120] have implemented the concept of functionally graded adhesive and found significant reductions in peak adhesive stresses around the free edges of the bond line. The adhesive joint was made of similar or dissimilar adherends and functionally modulus graded bond lines with smooth and continuous gradation function profiles. Results showed that peel and shear stress peaks in the functionally graded bond layer were much smaller and stress distributions along the joint region were more uniform than those of mono-modulus bond line adhesive joints. The same researchers indicated that the peel and shear strengths could be optimized by spatially controlling the adhesive modulus. However, their research is limited to only in-plane joint structures made of isotropic materials.

Dos Reis et al. [121] developed a numerical model for the different-graded distribution of adhesive properties along the overlap, using programmed step functions on finite element analysis (FEA) background to discretize and simplify the continuous properties distribution gradient. The same authors introduced cohesive zone modeling in the numerical model, enabling graded joint strength to predict effectively. The model was validated with experimental results of functionally graded joints available in the literature. The numerical model developed presents itself as a powerful tool to predict joint strength for functionally graded joints, without imposing large computational demands.

The structure having tee joint is expected to be subjected to tensile, compressive, bending, or combination during the service conditions. Stress analysis of these loading conditions is of utmost importance for a tee joint designer. Design and analysis of tee joint structure become a challenge for the designer/researcher when functionally graded adhesive materials and laminated FRP composites are used [122]. The performances and behavior of adhesively bonded double supported tee joint subjected to a general loading, including bending were analyzed. Three-dimensional geometrically nonlinear FE simulations have been carried out to evaluate the stresses at different surfaces under different loading conditions. Subsequently, a coupled stress failure criterion has been employed to evaluate the failure indices to predict damage onset location for the tee joint structure under general loading conditions. A series of numerical simulations have been performed to indicate the effect of functionally graded adhesive with varied modulus ratios on the out-of-plane stresses induced in the joint structure [122].

## 5.2 ANALYSIS AND EVALUATION OF DIFFERENT LOADINGS ON THE FAILURES OF TEE JOINT

Three-dimensional nonlinear FEAs have been carried out for a functionally graded adhesively bonded tee joint made of laminated FRP composites when the joint is subjected to different types of loadings [122]. The out-of-plane stress components have been evaluated along the interfacial surfaces of the bond line, which are subsequently used to predict the location of onset of failures in the joint. A series of numerical simulations have been performed to visualize functionally graded adhesive's effect on out-of-plane stress components [122].

The adhesively bonded tee joint shown in Figure 5.1 has been considered in the present analysis. The tee joint consists of a vertical plate, a base plate and two corner

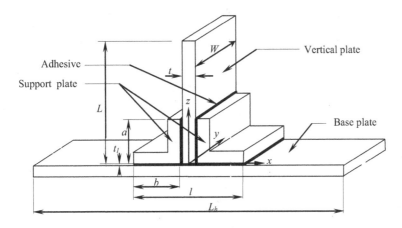

**FIGURE 5.1**  Double supported tee joint configuration.

supports, in which the vertical plate is bonded to the base plate with the help of two corner supports. Dimensions of the considered tee joint are given in Table 4.1. The vertical plate, support plates and base plate consist of $[0]_8$, graphite/epoxy FRP composite laminates. The thickness of each ply is taken as 0.25 mm. The behavior of FRP composites and adhesive are considered to be linear elastic in the present research work. The layer-wise orthotropic properties of FRP composite laminates shown in Table 3.2 are used for numerical simulation. Two types of adhesives, namely mono-modulus and functionally graded adhesives, bond the vertical plate with two support plates and base plate. The material properties of the mono-modulus adhesive are given in Table 4.3. The details of functionally graded adhesive considered in the present research are extensively explained in a proceeding section.

**Loading and constraints:** The tee joint with double support when subjected to different types of loading is considered for stress analysis. Different loading conditions such as tensile, compressive and bending are illustrated in Figure 5.2a–c, respectively, are applied to the tee joint.

A total load of 1.25 MPa has been applied on the top surface of the vertical plate to realize the different loading conditions. Both the ends of the base plate of the tee joint are fixed. This fixed boundary condition can be achieved by imposing suitable values on the displacement and slopes. More precisely referring to Figure 5.1, the boundary conditions imposed in the present FE simulation is expressed as below:

i. $x=-L_h/2;\ u_x=u_y=u_z=0\ \&\ \theta_x=\theta_y=\theta_z=0,$
ii. $x=L_h/2;\ u_x=u_y=u_z=0\ \&\ \theta_x=\theta_y=\theta_z=0.$

where $L_h$ is the total length of base plate; '$x$' measures the distance along the base plate; $u_x$, $u_y$ and $u_z$ represent the translational degrees of freedom along $x$, $y$ and $z$ axes, respectively; and $\theta_x$, $\theta_y$ and $\theta_z$ correspond to the respective slopes.

**FGA bond line:** Two bond lines, horizontal and vertical, are used to bond vertical plate with the base plate of the joint. Both bond lines are graded with functionally

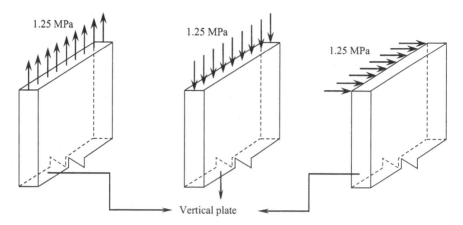

**FIGURE 5.2** Vertical plate of the double supported tee joint subjected to different loadings: (a) tensile, (b) compressive and (c) bending.

graded adhesive materials. Many researchers [93,96,107] considered linear and exponential function profiles for functionally graded materials. The same authors have used functionally graded adhesive materials in Chapter 4 (Section 4.2). One of the objectives was the optimization of the suitability of graded profile (linear or exponential) by comparing the magnitude of the stress components for tee joint structure. It was clearly spelt out that the linear material gradation function profile offers a better reduction in the magnitude of peak values of peel stress based on three-dimensional FEA. Again, the same authors have analyzed the effects of exponential and linear material gradation profiles on stress intensity factors (SIFs) in their earlier work [107,123]. It was observed that the SIF reductions are more pronounced for linear material gradation profiles compared to exponential profiles. In addition, Ravi Chandran and Barsoum [93] suggested the advantages of using functionally graded material with liner function profile based on the SIF value for a finite-width functionally graded plate with an embedded crack.

The reasons mentioned earlier justified using linear function profiles in the present research for improved structural performance. Because the structural performance in the case of functionally graded adhesively bonded tee joint largely depends on the peak vales of stress levels and SIF. Significant reductions in those values are observed due to the use of a linear graded function profile. Hence, the FGA is implemented through continuous and smooth variation of elastic modulus along the horizontal and vertical bond line, governed by linear function profile Eq. (3.1).

Because of the symmetry of the double supported tee joint, material gradation properties for the horizontal bond line are shown for half portion of the joint, shown in Figure 5.3a. As shown in Figure 5.3b, gradation properties along the vertical bond line are used for the present simulation when the tee joint is subjected to tensile and compressive loading. Figure 5.3c depicts the graded profile for vertical bond line in tee joint subjected to bending loading. Based on previous studies [93,109] and stress distributions, flexible adhesive with low elastic modulus values is used in the center of the tee joint and overlap ends of support plates. Whereas adhesive having the highest elastic modulus values is used at the center of the horizontal support plate. As shown in Figure 5.3a, Stiffer adhesive following linear profile has been used in between. For vertical bond, flexible adhesive is used near the base plate, and stiffer

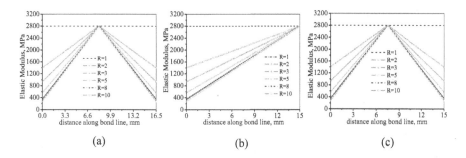

(a)                      (b)                      (c)

**FIGURE 5.3**  Gradation of elastic modulus along different bond layers with varied modulus ratio 'R': (a) horizontal bond layer, (b) vertical bond layer (tensile and compressive loading) and (c) vertical bond layer (bending load).

adhesive is used at the vertical overlap end in the tee joint when subjected to tensile and compressive loading, as shown in Figure 5.3b. However, the grading scheme is entirely different for the tee joint when subjected to bending load. Based on the stress analysis of joint subjected to bending, the flexible adhesive is used at both ends of vertical bond length. As shown in Figure 5.3c, Stiffer adhesive following linear profile has been used in the central region. In FE model, the changes of material property have been modeled discretely, by assigning each of the elements with the value of $E(x)$ at the middle of each of the elements [96].

**Meshing details:** Three-dimensional iso-parametric solid elements designated as SOLID 45 and SOLID 46 of ANSYS are used for meshing the adhesive bond layer and laminated FRP composite plates, respectively. One element is used to model the composite plate through the thickness, whereas two elements are used to model the adhesive layer. SOLID 46 is a layer version of the eight-node structural solid (SOLID 45) element, designed to model the laminated FRP composite plates ply by ply. Orthotropic material properties have been considered for each ply. These elements have eight nodes, and each node possesses three translational degrees of freedom. The mesh details used for modeling double supported tee joint structure are shown in Figure 5.4.

FE error estimation and convergence study have been conducted to achieve the optimized values of stresses. At the same time, validation of computational simulation is the primary method for building and quantifying confidence to establish an FE model with the available literature. That is why convergence study, error analysis, the establishment of results based on validation was done extensively by the same author [107] in their previous work. In the FE model, element size is considered (0.25 mm×0.25 mm×0.25 mm) for the adhesive layer. A mesh pattern of 132 elements (along the $x$-axis), 80 elements (along the $y$-axis) and 2 elements (along the $z$-axis)

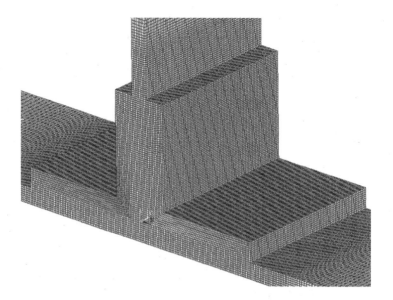

**FIGURE 5.4**   Meshing details for finite element model of double supported tee joint.

have been adopted to discretize the adhesive layer. Accordingly, the same mesh details have been incorporated in the present FE model of tee joint structure.

## 5.3  COHESIVE AND INTERFACIAL FAILURE ONSET IN TEE JOINT STRUCTURE

The mechanical strength of an adhesively bonded tee joint structure depends essentially on four parameters (i) the adhesion properties between adhesive and vertical/support/base plate materials, (ii) the cohesion properties of the adhesive materials, (iii) base and support plate of the joint with or without defects and (iv) the geometry and configuration of the joint. Understanding the fundamental nature of damage parameters is important in the reliability and safe design of adhesively bonded tee joints. Failure prediction of the joint is difficult because the failure strength and modes are different according to the various bonding methods, loading, geometrical and mechanical properties, etc. According to the studies, two kinds of failure modes are broadly in adhesively bonded tee joints. One is the failure of the adhesive layer, which includes interfacial failure called adhesion failure, and the other cohesion failure of adhesive. A cohesive failure philosophy formulates failure onset in the adhesive layer. Accordingly, Eq. (4.4) is used to predict the location of cohesive failure initiation in the tee joint. Tsai and Wu [60] coupled stress failure criterion is used to compute the interfacial failure indices 'e' over the horizontal and vertical bond line. In the tee joint structure, the interlaminar or out-of-plane stresses are responsible for the initiation of interfacial failures. Hence, these stress components only have been used in Eqs. (4.2) and (4.3) to evaluate the failure index values 'e'. Referring to Figure 5.1, for the vertical bond layer, the $x$-plane is out of the plane with respect to the global Cartesian coordinate system $(x, y, z)$ and accordingly, results are extracted. $\sigma_{xx}$, $\tau_{xy}$ and $\tau_{xz}$ are out-of-plane stresses evaluated along the vertical bond line interfacial surfaces. Thus, based on the magnitudes of failure indices 'e', the critical location for the onset of failure due to adhesion/cohesion can be identified.

## 5.4  INFLUENCE OF GRADED BOND LINE ON-RESISTANCE OF JOINT ASSEMBLY

A series of numerical simulations through FEA considering geometric nonlinearity are carried out for tee joint structure bonded with functionally graded adhesive. The out-of-plane stresses at the horizontal and vertical bond line interfaces have been evaluated when the tee joint is subjected to different loading such as tensile, compressive and bending. The failure indices 'e' using Eqs. (4.2)–(4.4) are evaluated for all the bond line surfaces of the tee joint. These failure indices have been used for finding out the location for the initiation of failures in the tee joint structure. Furthermore, efforts have been made to reduce out-of-plane stresses and ultimately failure indices by implementing functionally graded adhesive along horizontal and vertical bond lines. A series of numerical simulations are carried out in tee joint with functionally graded adhesive with varied modulus ratios ($R = 1, 2, 3, 5, 8, 10$).

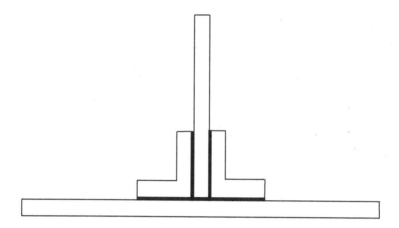

**FIGURE 5.5** Longitudinal cross-section through the central region of the tee joint along which the variations of out-of-plane stress distribution and failure indices have been presented.

Figure 5.5 shows the longitudinal cross-section through the central region of the tee joint. The variations of stress distribution and failure indices have been presented from the results of the three-dimensional FEA. The out-of-plane stress $\tau_{yz}$ along the considered cross-section is negligible and, hence, not considered and presented here. Accordingly, results are discussed for functionally graded adhesively bonded tee joint under varied loading conditions in proceeding sections. Horizontal bond line interfaces are deemed critical for tensile and compressive loading conditions, while vertical bond line interface is vulnerable under bending load [109]. Therefore, stress and failure index magnitudes are determined for horizontal bond line interface under tensile and compressive loading. In contrast, those are evaluated for the vertical and horizontal bond line of tee joints under bending load.

### 5.4.1  Tee Joint under Tensile Loading

The out-of-plane stresses are determined for three different bond line surfaces of horizontal bond line at the middle of tee joint subjected to tensile loading. The mid-surface of the adhesive layer is prone to cohesive failure due to both peel stress ($\sigma_{zz}$) and shear stress ($\tau_{xz}$). The peel and shear stress distributions along the midsurface of the functionally graded adhesive layer for varying modulus ratios ($R = 1, 2, 3, 5, 8, 10$) are shown in Figure 5.6a and b.

Modulus ratio '$R = 1$' refers to mono-modulus adhesive. Figure 5.6a shows that peel stress magnitudes are maximum at both the free ends of the adhesive layer and near the center of the joint. However, the highest magnitude of peel stress is observed at both the free ends of the adhesive layer. Referring to Figure 5.6b, it is clearly observed that the highest magnitude of out-of-plane shear stress ($\tau_{xz}$) occurs at both the free ends of the adhesive layer. This is because the geometric and material discontinuities give rise to the stress concentration effect at both the free edges of the adhesive layer. This behavior is in qualitative agreement with numerical and experimental results by Apalak et al. [109] and da Silva and Adams [110].

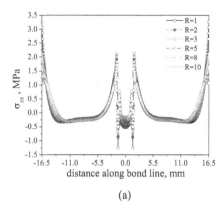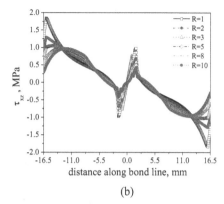

(a)                                                    (b)

**FIGURE 5.6** Out-of-plane stress distribution along the midsurface of horizontal bond line; (a) $\sigma_z$, (b) $\tau_{xz}$ with varied modulus ratios 'R' for graded tee joint under tensile loading.

Furthermore, the effects of graded adhesive on peel and shear stresses' magnitudes are clearly reflected in Figure 5.6a and b. Results indicate that when there is an increase in modulus ratios, peak peel ($\sigma_{zz}$) and out-of-plane shear stress ($\tau_{xz}$) decrease significantly. Figure 5.6a exhibits a 20%–70% reduction in peel stress level at both the free ends of adhesive layer by using graded adhesive with modulus ratios 'R' varying from 2 to 10. In the same line, out-of-plane shear stress ($\tau_{xz}$) significantly decreases at both the free ends and near the center of the joint. However, the reduction in magnitudes of peel stresses is not significant near the center of the joint.

Peel and out-of-plane shear stress distribution at the adhesive and base plate interface is depicted in Figure 5.7a and b. Referring to these figures, it is observed that reduction of intensities and trend of out-of-plane stresses is in line with that observed at the midplane of adhesive layer. However, the decline in magnitudes of shear stresses is insignificant near the center of the joint.

Figure 5.8a and b indicates peel and shear stress distribution along the adhesive and support plate interface. For mono-modulus adhesive ($R = 1$), the highest magnitude of peel stress is observed near the joint center, while the highest magnitude of out-of-plane shear stress is detected at both the free ends of the adhesive layer. Furthermore, a significant reduction in stress levels is observed using graded adhesive with modulus ratios ($R$) varying from 2 to 10.

The above-discussed results are in qualitative agreement with Kumar et al.'s numerical evidence [99,119]. Overall, the above results indicate that stress concentrations can be reduced by using graded adhesive along the bond line of tee joint under pull-off loading. It leads to improving the joint strength of the tee structure.

Knowledge of stress distributions on different surfaces, that is, (i) midsurface of adhesive, (ii) interfacial surface between the support plate and adhesive and (iii) interfacial surface between the base plate and adhesive layer of the joint, has been used to identify the location for failure initiation. The variations of failure indices 'e' along the different surfaces of tee joint with mono-modulus adhesive ($R = 1$) are shown in Figure 5.9a–c.

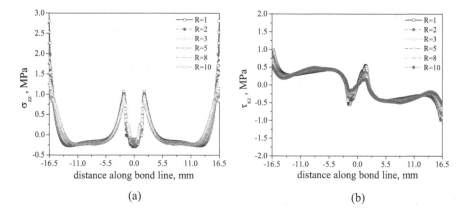

**FIGURE 5.7**   Out-of-plane stress distribution along the interface of base plate and horizontal adhesive layer; (a) $\sigma_{zz}$, (b) $\tau_{xz}$ with varied modulus ratios '$R$' for graded tee joint under tensile loading.

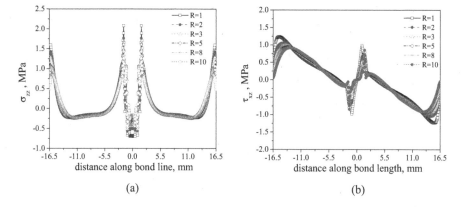

**FIGURE 5.8**   Out-of-plane stress distribution along the interface of support plate and horizontal adhesive layer; (a) $\sigma_z$, (b)$\tau_{xz}$ with varied modulus ratios '$R$' for graded tee joint under tensile loading.

It is clearly noticed from Figure 5.9 that failure onset takes place at the interface of the base plate and adhesive layer from both the free ends of horizontal bond length. Failure initiation location predicted in the present work is in full agreement with that observed by da Silva and Adams [110] and Apalak et al. [109]. The variations of failure indices along the above-predicted critical interfacial surface of the tee joint of functionally graded adhesive with modulus ratios($R$) varying from 2 to 10 are shown in Figure 5.10.

Results are compared between mono-modulus and graded adhesive. It shows a 20%–40% reduction in peak values of failure indices with graded adhesive. The present analysis speaks about the predicted location's reduced/delayed possibility of failure.

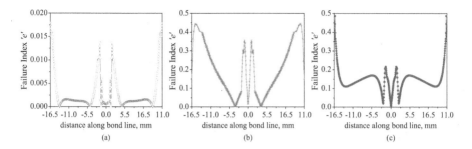

**FIGURE 5.9**  Variations of failure index '*e*' in tee joint with mono-modulus adhesive under tensile loading: (a) at the midsurface of adhesive layer, (b) at the interface of support plate and adhesive layer and (c) at the interface of base plate and adhesive layer.

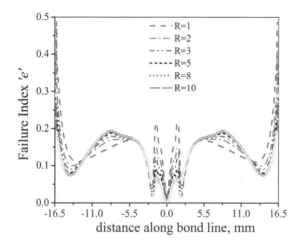

**FIGURE 5.10**  Variations of failure index '*e*' at the interface of base plate and adhesive layer of tee joint with functionally graded adhesive under tensile loading.

### 5.4.2  TEE JOINT UNDER COMPRESSIVE LOADING

The magnitudes of out-of-plane stresses are evaluated along three critical bond line (horizontal) surfaces of the tee joint subjected to compressive loading when functionally graded adhesive material is used. Figure 5.11a and b indicates the stress profile along the midsurface of the graded adhesive layer with modulus ratios '*R*' varying from 1 to 10.

The results show that peel and out-of-plane shear stress magnitudes are similar to those found in the case of joints subjected to tensile loading. However, the peel stress profile shows compressive nature for most joint regions. Referring to Figure 5.11, peak values of peel and shear stresses occur at both the free ends of bond length. Moreover, reduction in magnitudes of peel and shear stresses for the functionally modified bond line is found to be similar to that of the tensile loading situation. Figure 5.12a and b shows the magnitudes of stress distribution along the base plate

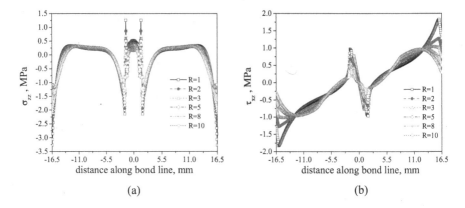

**FIGURE 5.11** Out-of-plane stress distribution along the midsurface of horizontal bond line; (a) $\sigma_z$, (b) $\tau_{xz}$ with varied modulus ratios '$R$' for graded tee joint under compressive loading.

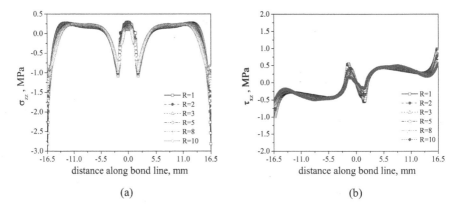

**FIGURE 5.12** Out-of-plane stress distribution along the interface of base plate and horizontal adhesive layer; (a) $\sigma_z$, (b) $\tau_{xz}$ with varied modulus ratios '$R$' for graded tee joint under compressive loading.

and adhesive layer interface, which is exactly similar to that occurred in the tee joint under tensile loading.

However, peel stresses are compressive in nature for most horizontal bond length regions. Above-discussed trend and nature of stresses are also identified at the interface of the support plate and adhesive layer, which is clearly depicted in Figure 5.13a and b. The above discussion is in qualitative agreement with results found by Apalak et al. [109].

The out-of-plane stress components have been used in Eqs. (4.2) and (4.3) to evaluate the failure index values along the horizontal bond line interfacial surfaces. As stress magnitudes are similar to that occurred in tensile loading, failure index profile and magnitudes remained the same in the joint under compressive loading, which is indicated in Figure 5.9a–c. The effects of graded adhesive having varied modulus ratios ($R = 1$–10) are also found to be similar to those that occurred in joints subjected

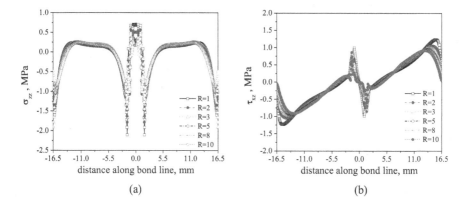

**FIGURE 5.13** Out-of-plane stress distribution along the interface of support plate and horizontal adhesive layer; (a) $\sigma_z$, (b) $\tau_{xz}$ with varied modulus ratios '*R*' for graded tee joint under compressive loading.

to tensile loading, which is shown in Figure 5.10. It indicates a delayed possibility of failure initiation from both the free ends of bond length.

### 5.4.3 Tee Joint under Bending Loading

From the physics of the problem, the vertical and horizontal bond line can be considered critical in the tee joint subjected to bending load. Hence, out-of-plane stress components are evaluated at all the interfacial surfaces of both the bond lines.

#### 5.4.3.1 Vertical Bond line

Out-of-plane stresses ($\sigma_{xx}$) and shear stress ($\tau_{xz}$) are determined on different surfaces, that is, (i) midsurface of adhesive layer, (ii) interfacial surface between the vertical plate and adhesive and (iii) interfacial surface between the vertical support plate and the adhesive layer of the joint. Here, *the x*-plane is out-of-plane for the vertical plate with respect to global Cartesian coordinate system (Figure 5.1). The distribution of the magnitudes and nature of out-of-plane normal ($\sigma_{xx}$) and shear stress ($\tau_{xz}$) along the midsurface of the left vertical adhesive layer are shown in Figure 5.14a and b.

For mono-modulus adhesive ($R = 1$), it is found that peak levels of out-of-plane normal and shear stress magnitudes are observed at both the ends of vertical bond length. However, the highest magnitudes of normal stress ($\sigma_{xx}$) are observed at the free end of the vertical adhesive layer. It is clearly observed that these peak stress levels along the vertical bond length are nearly 2.5–3 times higher than those that occurred at critical bond line surfaces of tee joint subjected to tensile/compressive loading.

Furthermore, referring to Figure 5.14a and b, the effects of functionally graded adhesive with modulus ratios ($R = 2, 3, 5, 8, 10$) are more pronounced at both the ends of vertical bond length, which is highly desired. 10%–50% reduction is observed in the peak values out-of-plane normal stresses, whereas 10%–80% reduction is seen for the decrease in the magnitudes of shear stresses by grading the vertical bond

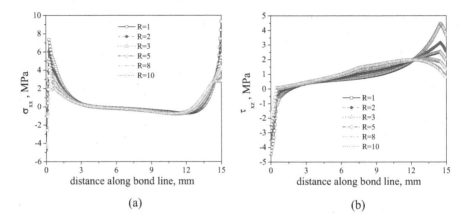

**FIGURE 5.14** Out-of-plane stress distribution along the midsurface of vertical left bond line; (a) $\sigma_z$, (b) $\tau_{xz}$ with varied modulus ratios '$R$' for graded tee joint under bending.

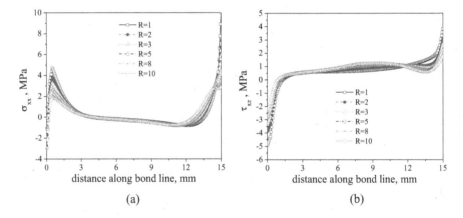

**FIGURE 5.15** Out-of-plane stress distribution along the interface of left vertical plate and adhesive layer; (a) $\sigma_z$, (b) $\tau_{xz}$ with varied modulus ratios '$R$' for graded tee joint under bending.

length with modulus ratios varying from 2 to 10. Similar reductions in the magnitudes of peak stress levels are also observed along the remaining two interfacial surfaces of the vertical bond line, i.e. vertical plate and adhesive layer (Figure 5.15) and vertical support plate and adhesive (Figure 5.16).

Overall, the functionally graded adhesive bond line offers an excellent reduction in the stress concentration at the critical regions, which is in qualitative agreement with the research work of by Kumar et al. [99,119]. The same results are also seen for the right vertical adhesive layers, which are evident due to symmetry.

The damage onset is predicted through failure indices '$e$' along the different interfacial surfaces of the left vertical bond line of joint with mono-modulus adhesive ($R = 1$), which are shown in Figure 5.17a–c.

It can be observed that failure initiation triggers from the vertical plate and adhesive interface and from the bottom location, which agrees very well with numerical

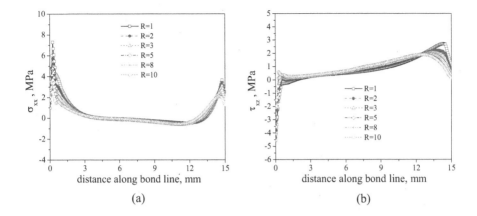

**FIGURE 5.16** Out-of-plane stress distribution along the interface of left vertical support plate and adhesive layer; (a) $\sigma_z$, (b) $\tau_{xz}$ with varied modulus ratios '$R$' for graded tee joint under bending.

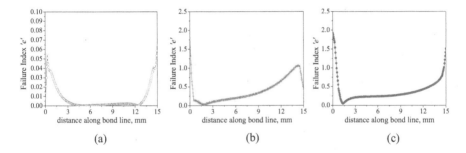

**FIGURE 5.17** Variations of failure index '$e$' along left vertical mono-modulus adhesive under bending load : (a) at the midsurface of adhesive layer, (b) at the interface of support plate and adhesive layer and (c) at the interface of vertical plate and adhesive layer.

evidence provided by Apalak et al. [109]. Under bending load, the peak value of failure index is four times higher than that occurred in the tee joint subjected to tensile/compressive load. Figure 5.18 shows the variations of failure indices '$e$' at the interface of the vertical plate and graded adhesive with varied modulus ratios ($R = 1$, 2, 3, 5, 8, 10). It is clearly seen that peak failure indices are reduced by 10%–50% compared with that of mono-modulus adhesive. The same results are also found for the right vertical bond line. This shows that graded adhesive exhibits excellent resistance to failure initiation from the critical region of bonded tee joint structure.

## 5.4.3.2 Horizontal Bond line

Based on the FEAs of the tee joint under bending load, interfacial out-of-plane stresses are evaluated along horizontal bond line surfaces. As discussed in previous sections, reductions in stress concentrations are also observed at the critical regions along the horizontal bond line with functionally graded adhesive. Based on the magnitudes of

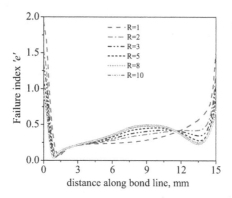

**FIGURE 5.18**   Variations of failure index '$e$' at the interface of left vertical plate and adhesive layer of tee joint with functionally graded adhesive under bending load.

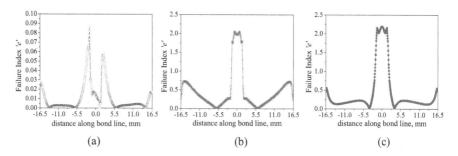

(a)                                        (b)                                        (c)

**FIGURE 5.19**   Variations of failure index '$e$' in tee joint with mono-modulus adhesive under bending load: (a) at the midsurface of adhesive layer, (b) at the interface of support plate and adhesive layer and (c) at the interface of base plate and adhesive layer.

out-of-plane stresses, failure indices are evaluated along the interfacial surfaces of the horizontal bond line and its variation is shown in Figure 5.19a–c.

The critical location for damage onset can be predicted, which is the base plate interface and adhesive and closed to the center of the joint. Referring to Figure 5.19c, the level of peak values of failure index is found to be four times higher than that observed in tee joint under tensile/compressive loading. The resistance to failure initiation from the predicted critical location is achieved by a horizontal graded bond line with varied modulus ratios, which can be visualized from Figure 5.20. However, reductions in the magnitudes of failure indices are less than that in the vertical bond line.

## 5.5   SUMMARY

Geometric nonlinear FEAs have been used to investigate the behavior of tee joints under varied loading conditions, which are used individually or in combination in actual applications. Out-of-plane stresses have been evaluated along the bond line surfaces to identify the critical region for failure onset. Furthermore, efforts are

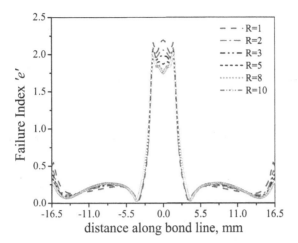

**FIGURE 5.20**  Variations of failure index '$e$' at the interface of base plate and adhesive layer of horizontal bond line of tee joint with functionally graded adhesive under bending load.

made to reduce peak stress levels by using functionally modified adhesive along the bond lines. The following specific conclusions are drawn based on results and observations made for the behavior of tee joint under varied loading.

- Peel and out-of-plane shear stress ($\tau_{xz}$) magnitudes are maximum at both the adhesive layer's free ends and near the joint's center subjected to tensile loading. However, the highest magnitude of peel stress is observed at both the free ends of the adhesive layer. But the reverse situation is observed for the horizontal support plate and adhesive interface.
- The magnitudes of peel and shear stresses along the horizontal bond line in the tee joint under compressive loading are similar to those found in the tee joint under tensile loading. However, the nature of peel stresses along bond length are compressive in the joint subjected to compressive loading.
- Out-of-plane stress levels along the bond line in the tee joint under bending load are 2.5–3 times higher than those observed in joints subjected to tensile/compressive loading. Therefore, bending is considered a critical loading condition for the tee joint structure.
- Under bending load, failure indices along the bond line are three to four times higher than those in the joint subjected to tensile/compressive loading.
- Under tensile/compressive loading, failure onset triggers from both the free ends of horizontal bond length and at the base plate and adhesive interface. However, the tee joint might fail under bending load from the two locations; one is the interface of the vertical plate and adhesive and from the bottom location. Another critical region is close to the center of joint and at the base plate and adhesive interface.
- Significant reduction in the peak peel and out-of-plane shear stresses is observed along all interfacial surfaces of the functionally graded bond line

of the tee joint under all loading conditions. The strength of the joint can be increased significantly.

- The failure indices along the critical bond line surfaces are significantly reduced by using functionally graded adhesive with varied modulus ratios. An excellent reduction (10%–50%) is observed for all loading conditions.
- Overall, the functionally graded bond line will retard the possibility of failure initiation by reducing peak levels of out-of-plane stresses by which the structural integrity of joint can be improved significantly. Hence, for a stronger and more efficient joint, a functional adhesive is recommended for a tee joint designer.

# 6 Effects of Functionally Graded Adhesive on Failures of Tubular Lap Joint of Laminated FRP Composites

## 6.1 INTRODUCTION

Application of FRP composites for piping systems was developed in response to significant corrosion and wear problems associated with metallic pipes in the chemical, pulp and paper, offshore oil and gas industries, etc. Pipes made from FRP composites have been widely used in wastewater treatment, power, petroleum, aerospace, automotive industries, etc. The complex layout of industrial piping systems, along with limitations associated with composite pipe manufacturing, demands repeatable and durable joining mechanisms. Mechanical joining methods of composite tubes such as trimming, bolting and fastening enhance stress raisers in the joint structure. Drawbacks associated with the mechanical joining of composite tubes lead to the present research toward adhesive bonding of tubular sections.

Adams and Peppiatt [124] performed stress analyses of adhesively bonded tubular joints under axial and torsional loading environments. It has been found that the finite element (FE) solutions were compared with the closed-form solutions. The same authors also indicated the effect of adhesive fillet and partial tapering of adherends on stress distributions in the adhesive layer. Thomsen [125] carried out a parametric study based on an elasto-static solution procedure for an adhesive bonded lap joint between two dissimilar orthotropic circular cylindrical laminated shells. In the past, the effects of various parameters such as an overlap length, adhesive layer and adherend stiffnesses on the stress distribution within the bond layer and adherends were studied extensively by many researchers. These parameters were optimized appropriately to maximize the load-bearing capability of the considered tubular joint. Pugno and Carpinteri [126] performed a general study on a single-lap tubular joint under axial loading to present the joint's static and dynamic structural behavior. They applied a fracture energy criterion to predict brittle crack propagation behavior in the joint. Esmaeel and Taheri [92,127] analyzed the effects of delamination with varied configurations on the peel and shear stress distributions induced in the adhesive layer of the composite tubular joint. Alwar and Nagaraja [128] carried out

DOI: 10.1201/9781003201113-6

a finite element analysis (FEA) of tubular lap joints subjected to axial loading, and the behavior of adhesive was assumed to be viscoelastic. Considerable reduction in magnitudes of stress peaks at the free edges of the adhesive layer was observed in the considered joint. Nemes et al. [129] used a variational method applied to the potential energy of deformation to analyze stresses in adhesively bonded cylindrical assemblies. They have investigated the effects of various geometrical and material parameters on the stress field in the joint. Whitcomb and Woo [64] conducted geometrically nonlinear FEAs to exhibit the potential for debond growth in adhesively bonded tubular joints. They highlighted the similarities and differences between tubular and plate joint behaviors. It was demonstrated that the damage driving forces in mode I ($G_I$) decreased with bond growth, but those forces in mode II ($G_{II}$) increased. Recently, Das and Pradhan [130] evaluated strain energy release rate (SERR) using virtual crack closure technique to assess the growth of adhesion failure in the tubular joint. They observed that adhesion failure damage propagation took place mainly in shearing mode.

Cognard et al. [131] used refined FE computations assuming materials of cylindrical joints as linearly elastic and illustrated stress concentrations in the joints under axial loading. In addition, the effects of different geometrical parameters such as adhesive thickness, bond length and tube thickness on the stress distribution were analyzed, and geometries that significantly limit the influence of edge effects were proposed. Many studies have been proposed to reduce the stress concentrations, especially in the case of lap joints, such as effects of spew and chamfer size [124,132] and influence of slots [133,134]. More recently, there are limited investigations to improve the joint strength by using functionally graded materials [116,135–137] and functionally graded adhesive (FGA) layer [89,99,108,119,138]. Apalak [135,136] performed three-dimensional stress analyses of adhesively bonded tubular lap joints with functionally graded adherends under tensile load and pressure loadings. Tubes were made of a functionally gradient layer between a ceramic and metal layers. It was noticed that continuous variation of material composition across the tube thickness plays a significant role in the peak values of the tube and adhesive stresses. The same author recommended a linear material function profile across the tube thickness to reduce the through-thickness stress levels. Kumar [119] made efforts to improve the strength of adhesively bonded tubular joints under axial load by reducing stress concentrations at the ends of overlap and distributing stresses uniformly over the bond layer. This strength improvement was achieved by using a FGA layer in the joint. The research was conducted by conducting axisymmetric elastic analysis. Results showed that peel and shear stress peaks were much smaller than those of mono-modulus adhesively bonded tubular joints. However, their research was limited to tubular joints made of isotropic adherends.

dos Reis et al. [139] studied the mechanical behavior of FGA joints loaded under impact conditions, using both experimental testing and numerical modeling. The results indicated that, unlike what is found for quasi-static loads, graded joints do not offer significant strength improvement under impact loads. In contrast, energy absorption is significantly increased. This behavior is explained by the completely different stress distribution on the adhesive layer for quasi-static and impact conditions, leading to the lower effectiveness of FGA joints under impact loads.

The structure having a tubular lap joint made of FRP composites is expected to be subjected to high axial load, internal pressure or a combination of both during the service conditions in chemical, offshore oil and gas industries, etc. Stress analysis pertaining to these loading conditions is utmost important for the designer. The joints are usually the weakest link in the system. Design and analysis of joint pipe structure becomes a challenge for the joint designer/researcher when FGA materials and laminated FRP composites are used. In the present investigation, the responses of adhesively bonded tubular lap joint structure subjected to a general loading were analyzed. Three-dimensional nonlinear FEAs considering geometric nonlinearity have been performed to compute the stresses at different surfaces under varied loading environments. Subsequently, a coupled stress failure criterion has been used to evaluate the failure indices to predict damage onset location for the considered joint structure under different loading conditions. The fracture mechanics principle is used to examine the sustainability of joint pipe structures having damage pre-existed at the critical locations. Moreover, the potential of FGA is explored along the bond layer in the joint pipe structure for improved damage growth resistance under a general loading environment. A series of numerical simulations have been conducted to establish the effect of graded adhesive with varied modulus ratios on damage propagation rate.

## 6.2 STRUCTURAL BEHAVIOR OF TUBULAR JOINT UNDER PRESSURE AND AXIAL LOADS

FEA can perform an accurate analysis of tubular single-lap joint, and this is a tool that can provide physical insight and accurate results. In the present research, SERR-based damage analyses of functionally graded adhesively bonded tubular lap joints of laminated fiber-reinforced polymeric (FRP) composites under varied loadings have been studied using three-dimensional geometrically nonlinear FEAs [58]. FE simulations have been carried out when a tubular joint is subjected to axial and pressure loadings.

**Modeling of tubular lap joint:** Tubular lap joint that consists of an adhesive layer and two composite tubes as shown in Figure 6.1, is considered for the present analysis. The isometric view showing the geometry and configurations of the adhesively bonded tubular lap joint is shown in Figure 6.1a, and the longitudinal sectional view is shown in Figure 6.1b. The main dimensions of the inner and outer tubes were taken as [130,140]: the inner radius of the inner tube $r_1 = 18.9$ mm, the inner radius of outer tube $r_2 = 20.05$ mm, the wall thickness of both tubes $t = 1$ mm, the length of each tube $l = 80$ mm, the adhesive thickness $\delta = 0.15$ mm, the overlap length $2c = 22$ mm and total length of joint structure $L = 138$ mm. The inner and outer tubes consist of $[0/90]_s$, graphite/epoxy composite laminates [130]. The thickness of each ply is considered as 0.25 mm.

In the present research work, the behavior of FRP composites and adhesives is considered to be linearly elastic. The layer-wise orthotropic material properties along with their strength values for FRP composite laminates are shown in Table 6.1. Two types of adhesives namely mono-modulus and FGAs are used to bond the inner tube with the outer tube. The material properties of the mono-modulus adhesive are given

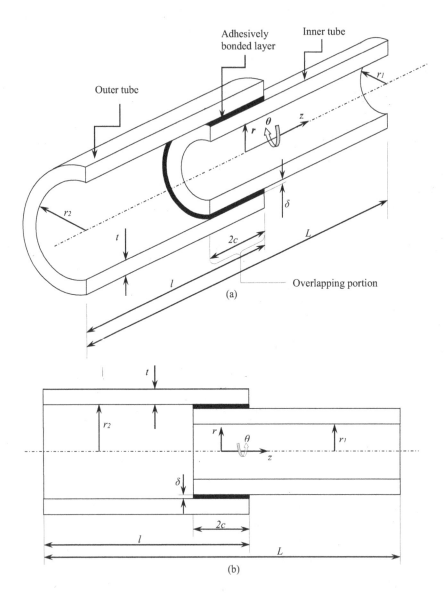

**FIGURE 6.1** Geometry and configuration of the adhesively bonded tubular lap joint: (a) Isometric view and (b) longitudinal sectional view.

in Table 4.3. The details of FGA considered in the present research are explained in the proceeding section.

**Loading and boundary conditions:** When subjected to pressure and axial loading, the adhesively bonded tubular lap joint structure is considered for stress analysis and damage propagation. The axial load is applied at the far end of the inner tube, equivalent to uniform loading of intensity of 10 MPa. An internal pressure of 10 MPa is applied to the internal surfaces of the inner and outer tubes [136]. To simulate the tubular joint under axial loading environment, the free edge of the outer tube is fixed. Tubular joint under pressure loading is simulated by fixing the free edges of the inner and outer tubes only in the longitudinal direction [136]. These fixed boundary conditions can be achieved by imposing suitable constraints in terms of displacement and slopes. Referring to Figure 6.1, the boundary conditions imposed in the present FE simulation for axial loadings are expressed as

$$\text{i.} \quad \text{for} \quad z = -L/2; \quad u = v = w = 0 \qquad (6.1)$$

and for pressure loading, following constraints are applied:

$$\text{ii.} \quad \text{for} \quad z = \pm L/2; \quad w = 0 \qquad (6.2)$$

where $L$ is the total length of joint structure. '$z$' measures the distance along the length of joint structure; $u$, $v$ indicates the radial, circumferential and axial displacements associated with the cylindrical coordinate system ($r$-$\theta$-$z$), respectively.

**Bond layer with FGA:** The linear function profile has been already justified in Section 5.2 for improved structural performance of the joint. Hence, in the present investigation, continuous smooth variation of elastic modulus of adhesive along the bond layer has been considered and implemented by applying several rings of adhesive of different moduli in the bond line, which is expressed by following linear function profile [58,93]:

$$E(z) = E_2 + (E_1 - E_2) \times \frac{z}{c} \quad \text{for} \quad (0 \leq z \leq c) \qquad (6.3)$$

$$E(z) = E_2 - (E_1 - E_2) \times \frac{z}{c} \quad \text{for} \quad (-c \leq z \leq 0) \qquad (6.4)$$

Material nonhomogeneity of graded bond layer have been evaluated in terms of modulus ratio ($R$), which is expressed as follows:

$$R = \frac{E_2}{E_1} \qquad (6.5)$$

The detail distribution of gradation properties with varied modulus ratios along the bond layer of the tubular lap joint structure is shown in Figure 6.2. Based on stress distributions [119,107], flexible adhesive having low values of elastic modulus ($E_1$) is used at both the overlapping end zones ($z = \pm c$). The stiffest adhesive with the highest elastic modulus values ($E_2$) is used at the central portion of the tubular joint structure ($z = 0$).

The upper bound modulus $E_2$ is taken as 2.8 GPa, and the lower bound modulus $E_1$ is varied according to modulus ratio '$R$' as expressed in Eq. (6.5). Stiffer adhesive following linear function profile as shown in Figure 6.2 has been used in the remaining bond layer. For modeling tubular joint with FGA, the adhesive region is assumed to have a Young's modulus changing along the $z$-axis using Eqs. (6.3) and (6.4). Infinite element model, the changes of material property have been modeled discretely by

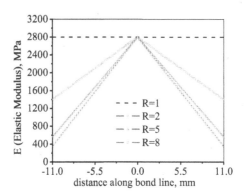

**FIGURE 6.2**   Gradation of elastic modulus ($E$) along bond length of the tubular lap joint for different modulus ratio '$R$'.

assigning the value of $E(z)$ at the middle for each of the elements [58,96] within the adhesive layer. The smooth variation of elastic moduli is ensured by keeping fine mesh (mesh size tends to zero) along the bond length. In FE model, element size is considered as (0.25 mm×0.347 mm×0.075 mm) for adhesive layer [58]. A mesh pattern of 360 elements (along circumferential direction), 88 elements (along axial direction) and 2 elements (along radial axis) have been adopted to discretize the adhesive layer. Nodes on interfacial surfaces are shared by both adhesive and tubes. Multipoint constraint (MPC) 184 elements have been used along the damage front using shared nodes to simulate the damage propagation.

**Meshing scheme:** The isotropic adhesive layer has been modeled using SOLID 45 brick elements. Two SOLID 45 elements are used to model the adhesive layer through the thickness (radial direction). On the other hand, isoparametric, 3D eight-node layer volume elements designated as SOLID 46 have been used to model the composite tube laminates ply-by-ply and orthotropic material properties have been considered for each ply (Table 6.1). One SOLID 46 element is used to model the composite tube through the thickness. These elements have eight nodes, and each node possesses three translational degrees of freedom. The mesh details used for modeling tubular joint structure are shown in Figure 6.3.

**Validation of FE model:** In the present FEA is validated by comparing the results for tubular lap joint under axial load with published literature [124]. Behavior of joint material was assumed to be linearly elastic. The magnitudes of peel stress ($\sigma_{rr}$) and shear stress ($\tau_{rz}$) in the adhesive layer of the joint are evaluated. These stress values have been nondimensionalized by dividing by the mean applied shear stress. Referring to Figure 6.4, peel and shear stress distribution within the midsurface of the adhesive layer of the joint shows good agreement with the available results [124].

## 6.3   DAMAGE GROWTH ANALYSES IN TUBULAR JOINT STRUCTURE

There are two vital modes of mechanical failure in tubular adhesively bonded joint structure: (i) adhesion/interfacial failure at the tube/adhesive interfaces and (ii)

**TABLE 6.1**

**Layer-wise Orthotropic Material Properties of Gr/E (T300/934) Composite [27, 29]**

Elastic properties

| | |
|---|---|
| $E_z$ | 127.50 (GPa) |
| $E_r$ | 4.80 (GPa) |
| $E_\theta$ | 9.00 (GPa) |
| $G_{zr} = G_{z\theta}$ | 4.80 (GPa) |
| $G_{\theta r}$ | 2.55 (GPa) |
| $v_{zr} = v_{z\theta}$ | 0.28 |
| $v_{\theta r}$ | 0.41 |
| Strengths | |
| $R_T$ (out-of-plane normal [transverse] strength) | 49 (MPa) |
| $S$ (out-of-plane shear strength) | 2.55 (MPa) |

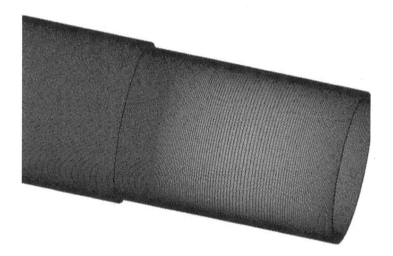

**FIGURE 6.3** Meshing scheme of finite element model for tubular lap joint.

cohesive failure within the adhesive bond layer. The interfacial failure takes place from the locations having stress concentration in tubular bonded joint structure. Tsai-Wu has identified the initiation of interfacial failure along the critical surfaces of bond layer coupled stress failure criterion [60] by computing failure indices, which takes into account the interaction of all six stress components given by

$$\frac{\sigma_r^2}{R_T^2} + \frac{\sigma_\theta^2}{\theta_T^2} + \frac{\sigma_z^2}{Z_T^2} + \frac{\tau_{r\theta}^2}{S_{r\theta}^2} + \frac{\tau_{rz}^2}{S_{rz}^2} + \frac{\tau_{z\theta}^2}{S_{z\theta}^2} + \sigma_r\left(\frac{1}{R_T} - \frac{1}{R_C}\right) + \sigma_\theta\left(\frac{1}{\theta_T} - \frac{1}{\theta_C}\right)$$

$$+ \sigma_z\left(\frac{1}{Z_T} - \frac{1}{Z_C}\right) + f_{r\theta}\sigma_\theta\sigma_r + f_{rz}\sigma_z\sigma_r + f_{z\theta}\sigma_z\sigma_\theta = e^2 \begin{cases} e \geq 1, & \text{failure} \\ e < 1, & \text{no failure} \end{cases} \quad (6.6)$$

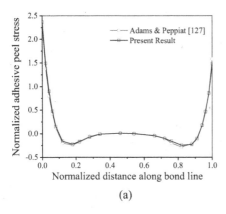
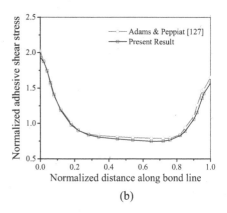

(a)                                                    (b)

**FIGURE 6.4** Validation of present result by comparing the distribution of out-of-plane stresses along the adhesive layer: (a) peel stress and (b) shear stress.

where $\sigma_r$, $\sigma_\theta$, $\sigma_z$, $\tau_{r\theta}$, $\tau_{rz}$ and $\tau_{z\theta}$ are the normal and shear stresses associated with the cylindrical coordinate system. $R_T$, $\theta_T$ and $Z_T$ are the allowable tensile strengths in the three principal material directions and $R_C$, $\theta_C$ and $Z_C$ are the allowable compressive strengths in the three principal material directions. $S_{r\theta}$, $S_{rz}$ and $S_{z\theta}$ are the shear strengths of the orthotropic layer in the various coupling modes. $f_{r\theta}$, $f_{rz}$ and $f_{z\theta}$ stand for coupling coefficients which reflects the interaction between $r$, $\theta$ and $z$ directions, respectively. In tubular joint structure, the interfacial failure is mainly attributed to the interlaminar stress effects, so only the interlaminar shear stresses ($\tau_{rz}$, $\tau_{r\theta}$) and through-the-thickness normal stress ($\sigma_r$) are required to be used to predict the initiation of interfacial failure. Therefore, the Tsai-Wu criterion as given in Eq. (6.6) can be simplified as

$$\left(\frac{\sigma_r}{R_T}\right)^2 + \left(\frac{\tau_{r\theta}}{S_{r\theta}}\right)^2 + \left(\frac{\tau_{rz}}{S_{rz}}\right)^2 = e^2 \tag{6.7}$$

where $R_T$ is the interlaminar normal strength and $S$ is the interlaminar shear strengths considered equal, that is, $S_{r\theta} = S_{rz} = S$. Similarly, a cohesive failure philosophy formulates the failure index of tubular joint in the adhesive layer. Parabolic yield criterion [61,97] used for prediction of the location of cohesive failure initiation in tubular joint structure is expressed as

$$(\sigma_1 - \sigma_2)^2 + (\sigma_2 - \sigma_3)^2 + (\sigma_3 - \sigma_1)^2 + 2(|Y_C| - Y_T)(\sigma_1 + \sigma_2 + \sigma_3) = 2e|Y_C|Y_T \tag{6.8}$$

where $\sigma_1$, $\sigma_2$ and $\sigma_3$ in the above equation represent the principal stresses and $Y_T$ and $Y_C$ indicates the yield strengths of adhesive in tension and compression, respectively. Failure indices for three surfaces of the bond layer can be evaluated by using Eqs. (6.7) and (6.8). The location for initiation of failure can be predicted based on the magnitudes of failure indices '$e$'. Furthermore, failure propagation in the tubular joint is analyzed by computing SERRs ($G_I$, $G_{II}$, $G_{III}$).

The damage growth behavior due to pre-embedded damages at the critical locations has been modeled. Their propagations are governed by the individual modes of SERR along the damage front for the considered tubular lap joint made of laminated FRP composite. In the laminated FRP composite tubes of the joint, due to their inherent complexities due to geometrical, loading and material properties, exact closed-form expressions for SERR are not possible. The singularity of crack-tip stress field in an orthotropic media is quite different from that of the conventional square-root singularity at the crack-tip inhomogeneous isotropic material system. This leads to evaluating interlaminar fracture energy released due to the propagation of the existing interfacial failure by a very small amount [108]. SERR procedure is suitable for assessing interfacial failure propagation behavior because it is based on a sound energy balanced principle implying its robustness, and also mode separation of SERR is possible. Irwin's [69] theory of crack closure has been followed for evaluating individual modes of SERPs. This aspect is very important as, in most cases, the fracture mechanism is a mixed-mode phenomenon in multidirectional laminated FRP composite tubes.

**Computation of SERR:** Based on magnitudes of failure indices values, the critical surface in the tubular joint has been identified. It is found that interfacial damage is expected to initiate from the inner tube and adhesive interface near the edge of bond layer closer to the loaded end of the tubular joint subjected to axial loading. The 2D representation of the tubular lap joint pre-embedded with interfacial damage is shown in Figure 6.5.

The SERR values are determined at any arbitrary point situating along the damage front. The damaged front considered in the present analysis is measured along the circumference of the tubular joint as shown in Figure 6.6. MPC elements are used along the interfacial failure front except over the damaged region.

The three components of SERR viz. $G_I$, $G_{II}$ and $G_{III}$ have been evaluated using MCCI and used as parameters for assessing the damage propagation characteristics [108,141]. Irwin's theory of crack closure can compute the SERRs along the damage front from those stresses and displacement fields. The strain energy released by the propagation of interfacial failure of length $a$ to $a+\Delta a$ is given by

$$W = \int_{a}^{a+\Delta a} \int_{-\frac{\Delta a}{2}}^{\frac{\Delta a}{2}} \sigma(z,\theta)\delta(z - \Delta a,\theta)dzd\theta \qquad (6.9)$$

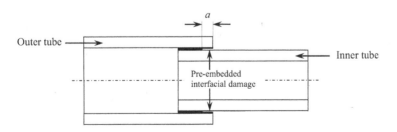

**FIGURE 6.5** Tubular lap joint pre-embedded with interfacial damage existing between the interface of the inner tube and adhesive layer.

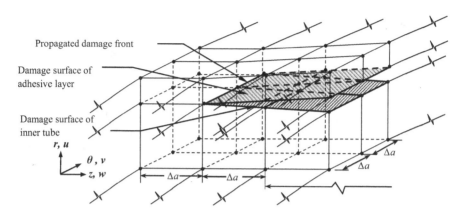

**FIGURE 6.6** MCCI applied to damage front located circumferentially along the tubular lap joint.

where $\delta(z - \Delta a, \theta)$ is the crack opening displacement between the top and bottom interfacial failure surfaces and $\sigma(z, \theta)$ is the stress at any point on the interfacial failure front required to close the separated area. Then the SERR ($G$) is obtained as

$$G = \lim_{\Delta a \to 0} \frac{W}{\Delta A} \tag{6.10}$$

where $\Delta A$ represents the interfacial failure propagated area and equals to one element area in $z - \theta$ plane, that is, $\Delta a \times \Delta a$ for the present case. The MCCI has the advantage of mode separation of SERR. This will help for a qualitative analysis of interfacial failure propagation behavior. Accordingly, the three components of SERRs $G_I$, $G_{II}$ and $G_{III}$ for Modes *I*, *II* and *III* can be shown as follows:

$$G_I = \lim_{\Delta a \to 0} \frac{1}{2\Delta A} \int_a^{a+\Delta a} \int_{-\frac{\Delta a}{2}}^{\frac{\Delta a}{2}} \sigma_r(z,\theta) \times [u_T(z - \Delta a, \theta) - u_B(z - \Delta a, \theta)] dz d\theta \tag{6.11}$$

$$G_{II} = \lim_{\Delta a \to 0} \frac{1}{2\Delta A} \int_a^{a+\Delta a} \int_{-\frac{\Delta a}{2}}^{\frac{\Delta a}{2}} \tau_{rz}(z,\theta) \times [w_T(z - \Delta a, \theta) - w_B(z - \Delta a, \theta)] dz d\theta \tag{6.12}$$

$$G_{III} = \lim_{\Delta a \to 0} \frac{1}{2\Delta A} \int_a^{a+\Delta a} \int_{-\frac{\Delta a}{2}}^{\frac{\Delta a}{2}} \tau_{r\theta}(z,\theta) \times (v_T(z - \Delta a, \theta) - v_B(z - \Delta a, \theta)] dz d\theta \tag{6.13}$$

**Damage initiation in the tubular joint under axial loading:** Out-of-plane stresses ($\sigma_r$, $\tau_{r\theta}$, $\tau_{rz}$) responsible for damage initiation are evaluated on different interfacial surfaces, that is, (i) inner tube and adhesive, (ii) midsurface of adhesive and (iii) outer tube and adhesive. Tsai-Wu coupled stress failure criterion using Eq. (6.7) has been used to compute failure indices 'e' on the interfacial surfaces, whereas parabolic yield criterion (Eq. 6.8) is used to evaluate the failure indices 'e' within the adhesive layer. The details of failure initiation and its propagation characteristics for tubular joints under axial loading environment are discussed here. The variations of failure indices along the critical surfaces of the bond layer of the tubular joint are shown in Figure 6.7. The edges of the bond layer are more vulnerable zones for interfacial failure initiation as failure indices for all the critical surfaces of the bond layer are the highest at these locations. Based on observation of magnitudes of failure indices for all the bond layer surfaces, it is noticed that the midsurface of adhesive is ignored for failure initiation compared to interfacial failure at inner/outer tube–adhesive interfaces.

Referring to Figure 6.7, it is observed that the magnitudes of failure index attains a peak value at the interface of the inner tube and adhesive layer near the edge of the bond layer closer to the loaded end of the joint structure. Hence, this location is more prone for the damage onset. This failure initiation location predicted in the present research work agrees well with the experimental evidence of Tong [54] under axial loading conditions.

**Damage growth in the tubular joint under axial loading with mono-modulus adhesive:** The damage growth analyses have been performed by simulating an embedded interfacial damage at the inner tube–adhesive interface. The three individual modes of SERR ($G_I$, $G_{II}$, $G_{III}$) have been evaluated circumferentially along the interfacial damage front. Results indicate that the magnitudes of 'G' components are constant over the damage front measured along the circumference. Hence, its visualization across the damage front is not presented in the present research work – the variations of individual modes of SERR for varied interfacial failure lengths 'a' are shown in Figure 6.8.

On comparing the individual modes of SERR, the contribution of $G_{III}$ is insignificant for interfacial propagation, which can be noticed from Figure 6.8c. Referring to Figure 6.8, results show that initially, $G_I$ and $G_{II}$ played a vital role in the propagation of interfacial damage. Hence, it can be said that, initially, interfacial damage propagates under mixed-mode conditions. $G_I$ plays an insignificant role in propagating interfacial damage, as shown in Figure 6.8a. For increased damage length,

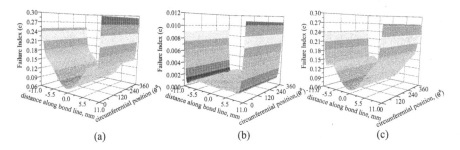

(a)　　　　　　　(b)　　　　　　　(c)

**FIGURE 6.7** Distribution of failure index 'e' in tubular lap joint under axial loading along the circumference: (a) at the interface of inner tube and adhesive layer, (b) the midsurface of adhesive layer and (c) at the interface of outer tube and adhesive layer.

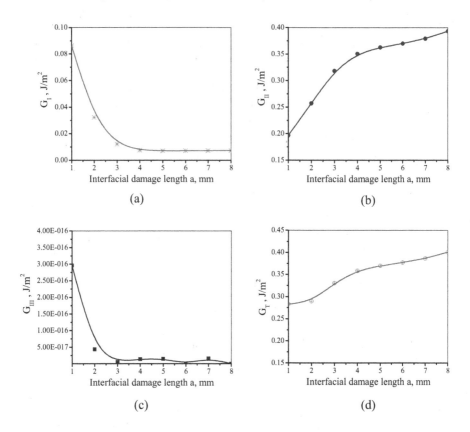

**FIGURE 6.8** Distributions of strain energy release rate components with varied interfacial damage length '$a$': (a) $G_I$, (b) $G_{II}$, (c) $G_{III}$ and (d) $G_T$ under axial loading condition.

interfacial damage propagates only due to SERR in mode II ($G_{II}$). Overall, it may be seen that the interfacial damage propagates under in-plane shearing mode in tubular lap joint structure when subjected to axial loading. The distribution of total SERR ($G_T$) is shown in Figure 6.8d, in which the significant contribution is due to $G_{II}$. Furthermore, it may be implied that the total SERR values increase with the propagation of damages. This means as the interfacial damage grows, the load-bearing capacity and structural integrity of the tubular joint reduces.

It is expected to improve the strength and lifetime of tubular lap joint for many structural applications by improving the bonding material's crack/damage growth resistance properties. Hence, on the basis of published literature [7,89,99,107,108,119,138], conventional isotropic adhesive along the bond layer of the tubular joint has been replaced by FGA. The influence of the functionally graded bond layer on the structural integrity of the tubular joint is discussed in the following section.

**Damage initiation in the tubular joint under pressure loading:** Figure 6.9 shows the distribution of failure indices '$e$' along the different surfaces of bond layer in tubular joint subjected to pressure loading. It is observed that the magnitudes of failure indices are compared to that when the joint is subjected to axial loading.

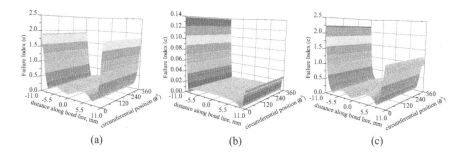

(a)                              (b)                              (c)

**FIGURE 6.9**  Variations of failure index '$e$' in tubular lap joint under pressure loading along the circumference: (a) at the interface of inner tube and adhesive layer, (b) the midsurface of adhesive layer and (c) at the interface of outer tube and adhesive layer.

This implies that the tubular joint under pressure loading is more critical than that of the axial loading environment. In pressure loading environment, failure index reaches the highest magnitude at the left free edge of the outer tube–adhesive interface. Hence, first, probable damage failure in the form of crack initiation is expected to trigger from this region, as evidenced by Apalak [136], when the joint functions in the same environment.

**Damage growth in the tubular joint under pressure loading with monomodulus adhesive:** Similar to axial loading, the magnitudes of '$G$' components are constant over the considered circumferential damage front in the tubular joint under pressure loading. Hence, its representation along the damage front is ignored. Results show that the contribution of $G_{III}$ is negligible, and thus its representation is not shown in the present research. As seen in Figure 6.10, mode I SERR ($G_I$) plays a significant role in damage propagation. The magnitude of $G_I$ grows continuously as damage propagates further.

The contribution of mode II SERR ($G_{II}$) is insignificant compared to $G_I$. Hence, opening mode is dominant for interfacial failure propagation in the joint under pressure loading. Moreover, it is observed that the magnitudes of SERR are significantly higher in pressure loading environments compared with axial loading conditions. Overall, the tubular joint under pressure loading is more prone to failure initiation and propagation.

## 6.4  EFFECT OF FGA ON INTERFACIAL FAILURE PROPAGATION

Interfacial damage growth in the tubular joint under axial loading has been analyzed by using FGA with varied modulus ratios '$R$' along the bond layer. It is assumed that the damages are pre-embedded at the critical location, which occurs at the inner tube and adhesive interface. The effects of graded adhesive are studied on interfacial damage propagation rate.

The significant effects of graded adhesive varying with modulus ratios $R = 1, 2, 5, 8$ are observed on the various components of SERR for varied interfacial failure lengths ($a = 1$–8 mm) when pre-embedded in the tubular joint structure are illustrated in Figure 6.11. It is noticed that SERR in mode I ($G_I$) reduces significantly with an

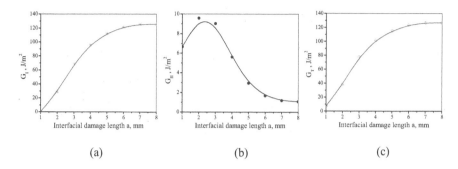

**FIGURE 6.10** Variations of strain energy release rate components with varied interfacial damage length '$a$': (a) $G_I$, (b) $G_{II}$ and (c) $G_T$ under pressure loading.

increase in modulus ratios '$R$'. As the damage propagates ($a = 1$–3 mm), the reduction in GI magnitudes is significant as depicted in Figure 6.11a. With further propagation of damage, reduction in magnitudes of $G_I$ diminishes. However, insignificant reductions in mode II SERR ($G_{II}$) for graded adhesive are also observed compared with mono-modulus adhesive ($R = 1$) as interfacial failure starts propagating ($a = 1$–3 mm). Furthermore, the differences of $G_{II}$ between the graded adhesive and mono-modulus adhesive become negligible as damage starts propagating. In a similar way, significant resistance to damage growth is also observed in respect of total SERR ($G_T$), which can be seen from Figure 6.11c. The behavior as mentioned earlier of tubular lap joint for interfacial damage growth resistance is in qualitative agreement with the observation made by Chandran and Barsoum [93] for crack growth analysis of functionally graded plates under pull-off loading conditions.

To visualize the reduction in total damage driving forces, Figure 6.12 shows the effect of material gradation of adhesive on total SERR ($G_T$) for tubular joints with varied interfacial damage lengths ($a$). Here, total SERR ($G_T$) is normalized by $G_T$, homo, the $G_T$ value for homogeneous adhesive material. These normalized values of total SERR are placed against the material nonhomogeneity parameter (modulus ratio, $R$). With an increase in modulus ratio ($R$), the difference between $G_T$ and $G_T$, homo increases. It is observed that $G_T$ values at damage front with nonhomogeneous adhesive decrease compared with that of homogeneous material except for higher interfacial lengths ($a = 7$–8 mm).

For any particular value of material nonhomogeneity($R$), the effect of the material gradation profile of adhesive on $G_T$ is more intense for shorter interfacial failure lengths ($a = 1$–3 mm). However, the highest reduction (23%) in damage driving forces is observed for the shortest interfacial failure length for modulus ratio, $R = 8$. The present profile of material gradation offers an excellent reduction in magnitudes of SERR for shorter interfacial failure lengths, which are the essential design characteristics of graded adhesive for improved interfacial failure growth resistance. The above-discussed behavior of tubular joint with graded adhesive is in qualitative agreement with many researchers [93,95,113].

Figure 6.13b shows the significant effects of graded adhesive varying with modulus ratios $R = 1, 2, 5, 8$ are observed on mode II SERR ($G_{II}$) for varied interfacial

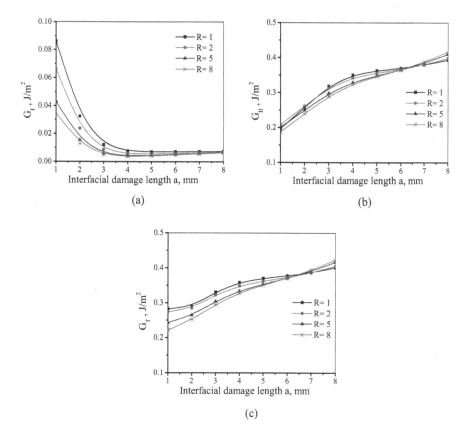

**FIGURE 6.11** Effects of functionally graded adhesive with varied modulus ratios 'R' on strain energy release rate components with varying interfacial damage length 'a': (a) $G_I$, (b) $G_{II}$ and (c) $G_T$ under axial loading condition.

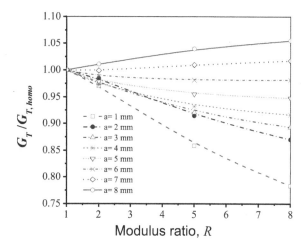

**FIGURE 6.12** Effect of different modulus ratios 'R' on total strain energy release rate for varied damage lengths in the joint under axial loading.

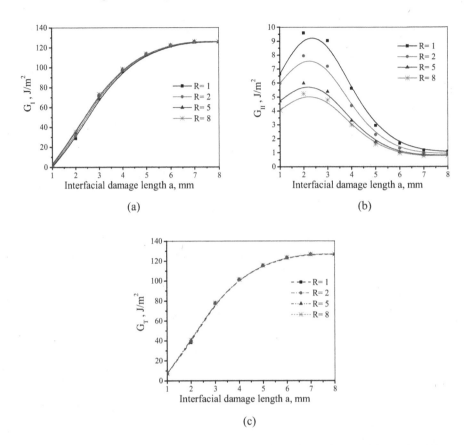

**FIGURE 6.13** Effects of functionally graded adhesive with varied modulus ratios '$R$' on strain energy release rate components in tubular lap joint under pressure load: (a) $G_I$, (b) $G_{II}$ and (c) $G_T$ under pressure loading.

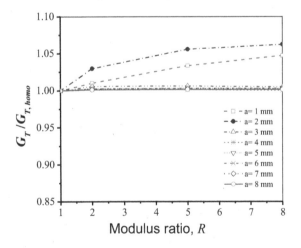

**FIGURE 6.14** Effect of different modulus ratios '$R$' on total strain energy release rate for varied damage lengths in the joint under pressure loading.

damage lengths in a tubular joint structure subjected to pressure loading. Substantial reductions in the magnitudes of $G_{II}$ are observed with increasing modulus ratios. However, reductions in SERR in mode I ($G_I$) and total SERR ($G_T$) are insignificant with an increase in modulus ratio as shown in Figure 6.13a and c.

Instead, magnitudes of total SERR ($G_T$) for varied damage length with graded adhesive are slightly increasing with modulus ratios. However, increase in SERR is prominent for shorter damage lengths (Figure 6.14), which is not desirable. Overall, the tubular joint under pressure loading does not offer any additional damage growth resistance using FGA.

## 6.5  SUMMARY

Interfacial damage growth in functionally graded adhesively bonded tubular lap joints with composites is a subject of significant importance and essential to designers. 3D geometrically nonlinear FEAs have been used to model the interfacial damage growth that existed in tubular joints under varied loading environments. The fracture mechanics-based approach and MCCI are used to compute the fracture parameter SERR to study damage growth. The following specific findings have been noticed from this research, which will provide important information while designing a tubular joint structure with graded adhesive:

- The interfacial damage growth rate is significantly higher in the tubular joint under pressure than tensile loading. Hence, it can be said that the tubular joint is vulnerable to failure and, the failure will propagate at a faster rate when subjected to pressure loads.
- In-pane shear mode SERR is responsible for damage propagation in the tubular joint under axial load, whereas opening mode SERR plays a vital role for damage propagation when the joint is subjected to pressure loading.
- Under axial loading conditions, significant interfacial damage growth resistance is observed in the tubular lap joint with FGA compared with mono-modulus adhesive. Moreover, the effect of material gradation of adhesive on total SERR is more pronounced for shorter interfacial failure lengths. However, same is not true for the tubular joint when subjected to pressure loading.
- Overall, tubular joint with FGA layer offers improved interfacial damage growth resistance, enhancing the structural integrity and service life of the tubular joint structure.

# 7 Design and Analyses of Tubular Socket Joints with Functionally Graded Adhesive Bond

## 7.1 INTRODUCTION

Adhesive bonding is the most attractive connection method in composite pipe joints because it can effectively lower the stress concentration through smoother load transfer between the connecting members. In addition, adhesive bonds are generally corrosion-free compared with mechanical fasteners. The most commonly used joining methods for composite pipes are as follows: (i) adhesive-bonded socket joints, (ii) butt-and-strap joints, (iii) heat-activated coupling joints and (iv) flanged joints. Most composite flanges are connected to composite pipes with the first three permanent joining methods. The basic configuration adopted in all permanent joints is that there are two pieces of composite pipes to be joined, a coupling to carry the load at the connection and a medium to transfer the load from the pipe to the coupling. Therefore, a general adhesive-bonded tubular socket joint analysis can be used to understand the criticalities involved in the three types of permanent composite pipe joints cited above.

This chapter is devoted to developing an FE-based simulation technique to study the onset and growth of adhesion/interfacial failures in functionally graded bonded tubular socket joints. The developed FE model provides sufficient scope for a detailed design and analysis of the bonded tubular socket joint in terms of stress analysis within the joint, studying the interfacial failure growth characteristics, and investigating the effect of graded adhesive on the growth of interfacial failures. The strain energy release rate (SERR) components calculated using modified crack closure integral (MCCI) vis-à-vis virtual crack closure technique (VCCT) have been used as the characterizing parameters for assessing the growth of interfacial failures in the present analyses.

## 7.2 STUDY OF INTERFACIAL FAILURES IN TUBULAR SOCKET JOINT

Computational modeling often can eliminate costly experiments and provide more information that can be obtained experimentally. Computational modeling has played an essential role in FGM research to date. Because of the considerable complexity

DOI: 10.1201/9781003201113-7

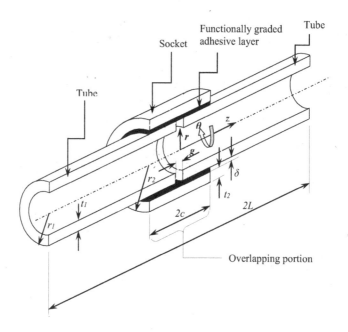

**FIGURE 7.1**    Adhesively bonded tubular socket joint: geometry and configuration.

involved is expected to play an even more significant role in future developments. In the present research, three-dimensional geometrically nonlinear FE analyses have been used to model the interfacial failure growth that pre-existed in tubular socket joints under an axial loading environment. The fracture mechanics-based approach and MCCI are employed to access the insight details on the structural behavior of the joint. A series of numerical simulations have been carried out to explore the potential influence of functionally graded bond layers on the structural integrity of the joint.

**Modeling of the tubular socket joint:** The geometry and configuration of bonded tubular socket joint analyzed is shown in Figure 7.1. The socket and both the tubes of the tubular socket joint are made of Gr/E laminated FRP composites (T300/934) with ply configuration $[0/90]_s$.

The material properties and their strength values for tubes/sockets and adhesive are given in Tables 6.1 and 4.3. The adhesive thickness ($\delta = 0.1$ mm), total length of the structure ($2L = 178$ mm), outer radius of tube ($r_1 = 14.4$ mm), outer radius of socket ($r_2 = 16$ mm), tube thickness ($t_1 = 1$ mm), coupling thickness ($t_2 = 1.5$ mm), coupling length ($2c = 26$ mm) and gap between the tubes ($g = 0.2$ mm) have been adopted from literature [140,141]. The adhesively bonded tubular socket joint structure subjected to axial loading is considered for stress analysis and damage propagation. Two types of adhesives, namely mono-modulus and functionally graded adhesives, are used to bond both the tubes with sockets. The material properties of mono-modulus adhesive and gradation profile for the functionally graded bond line are described in Section 7.3.

The free edge of one of the tubes is fixed, and tensile load is applied at the far end of another tube, equivalent to a uniform loading of 10 MPa to simulate tubular joint under an axial loading environment. Referring to Figure 7.1, restrained boundary conditions imposed in the present FE simulation for tubular socket joint under axial loadings are expressed as

$$\text{for} \quad z = -L; u = v = w = 0 \tag{7.1}$$

where '$2L$' is the total length of joint structure and '$z$' measures the distance along the length of joint structure. The parameters $u$, $v$, and $w$ indicates the radial, circumferential and axial displacements associated with the cylindrical coordinate system ($r$-$\theta$-$z$), respectively.

## 7.3 MATERIAL-TAILORED BOND LINE FOR IMPROVED JOINT STRUCTURAL INTEGRITY

The continuous and smooth linear variation of elastic modulus of adhesive along the bond layer has been considered and employed by applying several rings of adhesive of different moduli in the bond line by Eqs. (6.3) and (6.4). Because of the symmetry of the tubular socket joint, the detailed distribution of gradation properties with varied modulus ratios for the bond line is shown for half portion of the joint, which is exhibited in Figure 7.2.

Based on stress distributions of the present research problem and from the literature study, flexible adhesive having low values of elastic modulus ($E_1$) is used at both the overlapping end zones ($z = \pm c$) and at the center of the joint ($z = 0$). The stiffest adhesive with the highest elastic modulus values ($E_2$) is used at location, $z = \pm c/2$ of the tubular socket joint structure. In FE model, the changes of material property have been modeled discretely, by assigning each of the elements the value of $E(z)$ at the middle of each of the elements [59,96]. The continuous and smooth variation of elastic modulus along the bondline is ensured by adopting fine mesh for the considered FE model. Nodes on interfacial surfaces are shared by both adhesive and tubes/sockets [59]. Multi-Point Constraint (MPC) 184 elements have been used along the damage front using shared nodes to simulate the damage propagation.

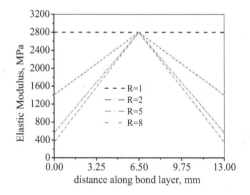

**FIGURE 7.2** Gradation of elastic modulus ($E$) along bond length of the tubular socket joint for different modulus ratio '$R$'.

The present work has constructed the FEM model of tubular socket joint using three-dimensional brick elements. Literature [142–144] shows that the three-dimensional brick element models are more accurate specifically in separating the total SERR into individual components $G_I$, $G_{II}$ and $G_{III}$. However, modeling and computational effort may become prohibitively large by using many layers of brick elements through the thickness to model the individual plies of laminated FRP composite tubes. Therefore, it is required to use layered volume elements to enhance computational efficiency without compromising the accuracy of the FE analysis for the considered tubular socket joint structure.

In the present FE analysis, isoparametric, three-dimensional eight-node layer volume elements designated as SOLID 46 have been used to model the composite tube laminates ply-by-ply and orthotropic material properties have been considered for each ply. The isotropic adhesive layer has been modeled using SOLID 45 brick elements. These elements have eight nodes, and each node possesses three translational degrees of freedom. The varied mesh density appropriately used for discretizing the tubular joint structure is illustrated in Figure 7.3.

**Onset and growth of interfacial failures in tubular socket joint:** Bonded tubular socket joint can be failed through various modes, viz. (i) interfacial failure at the tube/socket-adhesive interfaces due to excessive peel and shear stresses, (ii) cohesion failure within the adhesive layer, and (iii) interlaminar failures/delaminations in tubes/socket caused by interlaminar stresses. Cohesion and interlaminar delamination failure modes pertaining specifically to the failure of adhesive-bonded joints are discussed in detail by Dattaguru et al. [145] and Sheppard et al. [146]. The present research is focused on the interfacial/adhesion failure of tubular socket joints caused by interfacial peel and shear stresses.

Under three-dimensional stress states in the coupling region, the initiation of interfacial failures at the tubes/socket-adhesive interfaces generally can be evaluated by the Tsai-Wu coupled stress criterion given by Eq. (6.7). Failure indices calculated based on these criteria revealed that the possibility of interfacial failure initiation would be from the free edges of the tube–adhesive interfaces of the bonded tubular socket joint under axial loading. A detailed discussion in this regard has been given in the next section. The growth of adhesion failure in the tubular socket joint is performed by simulating pre-embedded interfacial failures near the free edges of

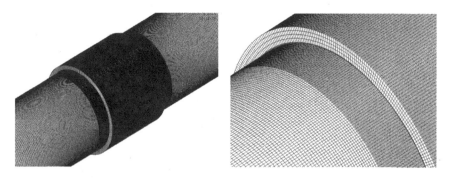

**FIGURE 7.3**  Mesh details for tubular socket joint: (a) full model and (b) zoomed view.

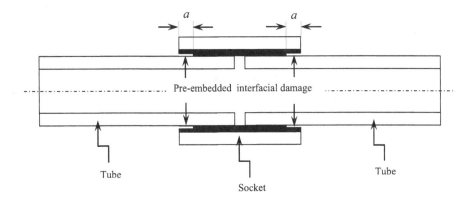

**FIGURE 7.4** Tubular socket joint pre-embedded with interfacial damage existing between the interface of the tube and adhesive layer.

the coupling region, as illustrated in Figure 7.4. The SERR, which is one of the key parameters responsible for the propagation of interfacial failure, is computed using the MCCI vis-à-vis VCCT (Eqs. 6.11–6.13).

**Results and discussion:** Geometrically nonlinear 3D finite element analyses have been carried out for tubular socket joints made of laminated FRP composites under axial loading. Out-of-plane stresses ($\sigma_r$, $\tau_{r\theta}$, $\tau_{rz}$) responsible for damage initiation are evaluated on different interfacial surfaces, that is, (i) tubes and adhesive, (ii) midsurface of adhesive and (iii) socket and adhesive. Tsai-Wu coupled stress failure criterion using Eq. (6.7) has been used to compute failure indices 'e' on the interfacial surfaces. In contrast, the parabolic yield criterion (Eq. 6.8) is used to evaluate the failure indices 'e' within the adhesive layer. The details of failure initiation and its propagation characteristics for tubular socket joint are discussed below:

The distributions of failure indices along the critical surfaces (tube/socket-adhesive, adhesive layer) of the bond line of tubular socket joint are illustrated in Figure 7.5. Free edges of the coupling region are more vulnerable zones for interfacial failure initiation as failure indices for all the critical surfaces of the bond layer are the highest at these locations.

Referring to Figure 7.5, it is seen that the midsurface of the bond line is a safer zone from the cohesion failure point of view compared with other critical interfaces. It is observed from Figure 7.5 that the magnitudes of failure index attain a peak value at the free edges of tube–adhesive interfaces in the coupling region of bonded tubular socket joint structure. Hence, this zone is more prone the damage onset. This failure initiation location predicted in the present research work agrees well with the experimental evidence of Tong [54] under axial loading conditions.

**Effect of functionally graded adhesive on the onset of interfacial failure:** The distributions of failure indices along the above predicted critical interfacial surface of the tubular socket joint of functionally graded adhesive with modulus ratios ($R$) varying from 1 to 8 are shown in Figure 7.6. Results are compared between monomodulus and graded adhesive. Significant reduction (20%–50%) in peak values of failure indices is noticed near the free edges of both the tube–adhesive interfaces in the coupling region, using a functionally graded bond layer with modulus ratios ($R$)

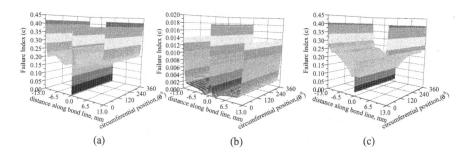

FIGURE 7.5   Distributions of failure index '*e*' in tubular socket joint: (a) at the interface of the tube and adhesive layer, (b) the midsurface of adhesive layer and (c) at the interface of socket and adhesive layer.

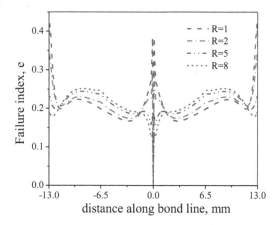

FIGURE 7.6   Variations of failure index '*e*' at the interface of the tube and adhesive layer of the tubular socket joint with functionally graded adhesive.

varying from 2 to 8. Considerable reduction is also noticed from the center of the joint region. This behavior of the graded tubular socket joint is in qualitative agreement with numerical evidence found by Kumar [99,119]. The present analysis concerns the reduced/delayed possibility of failure initiation from the predicted location.

**Interfacial failure growth in tubular joint with mono-modulus adhesive:** The damage growth analyses have been performed by simulating pre-embedded interfacial damage at the tube–adhesive interface. Figure 7.7a and b shows the variations of different modes of SERR ($G_I$, $G_{II}$, $G_{III}$) along the circumferential damage fronts with simultaneous propagation of interfacial failures from both the free edges of the coupling region (near the clamped edge and loaded edge).

The different modes of SERR for both interfacial failures are similar in terms of magnitude and trend of variations. In other words, it means that interfacial failures initiating from free edges of the tube–adhesive interfaces of tubular socket joint would propagate at the same rate. Therefore, visualization of SERR at the damage front from one of the free edges of the coupling region has been represented in the present research work. It is also observed from Figure 7.7 that the magnitudes of '*G*'

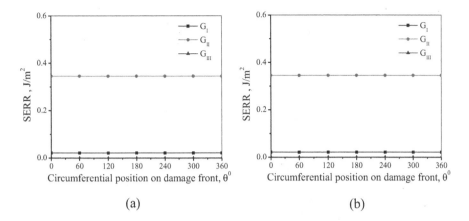

**FIGURE 7.7** Variations of different modes of strain energy release rate along the circumference of damage fronts with interfacial failure length, $a = 2$ mm near (a) the clamped edge and (b) the loaded edge of the coupling length of the bonded tubular socket joint.

components are found to be constant over the damage fronts measured along the circumference. Hence, its representation across the damage front is ignored in the present research work. This evidence shows self-similar growth of adhesion failures from the free edges of tube–adhesive interfaces. The variations of individual modes of SERR for varied interfacial failure lengths '$a$' are illustrated in Figure 7.8.

On comparing the individual modes of SERR, the contribution of $G_{III}$ is insignificant for interfacial propagation, which can be noticed in Figure 7.8c. Referring to Figure 7.8, results show that initially, $G_I$ and $G_{II}$ played a vital role in the propagation of interfacial damage. Hence, it can be said that initially, interfacial damage propagates under mixed-mode conditions. With an increase in damage length, $G_I$ plays an insignificant role in further propagation of interfacial damage, shown in Figure 7.8a. Interfacial damage propagates only due to SERR in mode II ($G_{II}$) for increased damage length. Overall, it may be seen that the interfacial damage propagates under in-plane shearing mode in tubular socket joint structure under axial loading. The distribution of total SERR ($G_T$) is shown in Figure 7.8d in which the significant contribution is due to shear mode ($G_{II}$). Furthermore, it may be noticed that the total SERR values increase with the propagation of damages. This means the growth interfacial damage affects the structural integrity of the tubular joint socket joint.

It is essential to improve the strength and lifetime of the tubular socket joint for many structural applications by improving the bonding material's crack/damage growth resistance properties. Hence, based on published literature [7,89,99,107,119,138], conventional isotropic adhesive along the bond layer of the tubular joint has been replaced by functionally graded adhesive. The influence of the functionally graded bond layer on the structural integrity of the tubular joint is discussed in the following section.

**Effect of functionally graded adhesive on interfacial failure growth in tubular socket joint:** Interfacial failure growth in the tubular socket joint under axial loading has been analyzed by using functionally graded adhesive with varied modulus ratios '$R$' along with the bond layer. It is assumed that the damages are pre-embedded

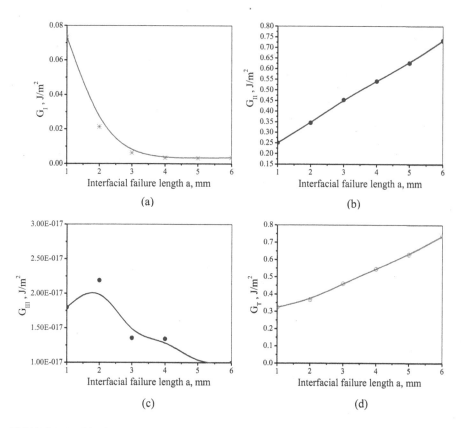

**FIGURE 7.8**  Distributions of strain energy release rate components with varied interfacial damage length '$a$': (a) $G_I$, (b) $G_{II}$, (c) $G_{III}$ and (d) $G_T$ under axial loading condition.

at the critical location which happens to be occurred from the free edges of the tube–adhesive interfaces of the coupling region in the tubular socket joint. The influence of graded adhesive is demonstrated on interfacial failure propagation rate. The significant effects of graded adhesive varying with modulus ratios $R = 1, 2, 5, 8$ are observed on the various components of SERR for varied interfacial failure lengths ($a = 1$ to $6\,\text{mm}$) when pre-embedded in tubular socket joint structure, as illustrated in Figure 7.9.

It is noticed that SERR in mode I ($G_I$) reduces significantly with an increase in modulus ratios '$R$'. Initially, as the damage propagates ($a = 1$–$3\,\text{mm}$), reduction in magnitudes of $G_I$ is significant, as depicted in Figure 7.9a. However, a rise in $G_I$ has been noticed beyond interfacial failure length ($a = 3\,\text{mm}$).

Similarly, considerable reductions in mode II SERR ($G_{II}$) for graded adhesive are also observed compared with mono-modulus adhesive ($R = 1$) as interfacial failure starts propagating ($a = 1$–$3\,\text{mm}$). Similarly, significant resistance to damage growth is also observed in respect to total SERR ($G_T$), as shown in Figure 7.9c. Aforesaid improved interfacial failure growth resistance behavior of tubular socket joint is in qualitative agreement with the observation made by Chandran and Barsoum [93] for crack growth analysis of functionally graded plates under pull-off loading condition.

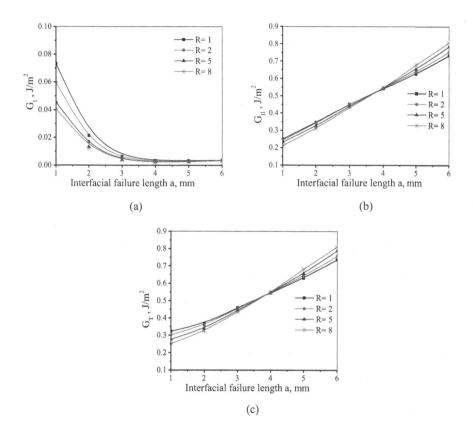

**FIGURE 7.9** Effects of functionally graded adhesive with varied modulus ratios '$R$' on strain energy release rate components with varying interfacial failure length '$a$': (a) $G_I$, (b) $G_{II}$ and (c) $G_T$.

The exact reduction in magnitudes of total SERR at interfacial failure front due to the use of functionally graded adhesive with different modulus ratios is clearly illustrated in Figure 7.10.

Here total SERR ($G_T$) is normalized by $G_{T,\ homo}$, the $G_T$ value for mono-modulus adhesive. These normalized values of total SERR are placed against the material nonhomogeneity parameter (modulus ratio, $R$). With the increase in modulus ratio ($R$), the difference between $G_T$ and $G_{T,\ homo}$ increases. It is observed that $G_T$ values at the interfacial failure front with a nonhomogeneous adhesive decrease compared with that of homogeneous material except for higher interfacial lengths ($a = 4$–$6$ mm).

For any particular value of material nonhomogeneity($R$), the effect of the material gradation profile of adhesive on $G_T$ is more intense for shorter interfacial failure lengths ($a = 1$–$3$ mm). However, the highest reduction (25%) in damage driving forces is observed for the shortest interfacial failure length for modulus ratio, $R = 8$. Considered linear material gradation of bond layer indicates an excellent reduction in magnitudes of SERR for shorter interfacial failure lengths, which are the desired design characteristics of graded adhesive for improved interfacial failure growth

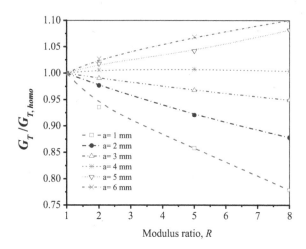

**FIGURE 7.10**   Effect of different modulus ratios '$R$' on total strain energy release rate for varied damage lengths in tubular socket joint.

resistance. This behavior of tubular socket joint with graded adhesive is in qualitative agreement with many researchers [93,95].

## 7.4   SUMMARY

This research presents interfacial failure analyses of laminated FRP composite tubular socket joints to identify the critical locations for damage onset. Furthermore, interfacial failure propagation has been extensively studied by 3D FE analyses with varied pre-existed interfacial failure lengths. The individual modes of SERR along the damage front have been evaluated using fracture mechanics-based approach. The potential use of functionally graded adhesive has been explored in tubular socket joints to reduce damage growth driving forces. The conclusions, based on the detailed results and observations of the structural behavior of functionally graded bonded tubular socket joint, are as follows:

- In-pane shear mode SERR is responsible for damage propagation in tubular socket joint under axial load.
- Under axial loading conditions, significant interfacial damage growth resistance is observed in the tubular socket joint with functionally graded adhesive compared with mono-modulus adhesive. Moreover, the effect of material gradation of adhesive on total SERR is more pronounced for shorter interfacial failure lengths.
- Overall, high damage growth resistance will retard propagation, hence improving the tubular socket joint's service life. Therefore, recommendations can be made to use a functionally graded bond layer in the tubular socket joint.

# References

[1] Castagnetti D, Spaggiari A, Dragoni E. Robust shape optimization of tubular butt joints for characterizing thin adhesive layers under uniform normal and shear stresses. *Journal of Adhesion Science and Technology* 2010; 24: 1959–1976.

[2] Spaggiari A, Castagnetti D, Dragoni E. Experimental tests on tubular bonded butt specimens: Effect of relief grooves on tensile strength of the adhesive. *Journal of Adhesion* 2012; 88: 499–512.

[3] Cognard JY. Numerical analysis of edge effects in adhesively-bonded assemblies application to the determination of the adhesive behavior. *Computers and Structures* 2008; 86: 1704–1717.

[4] Zhao X, Adams RD, da Silva LFM. Single lap joints with rounded adherend corners: Experimental results and strength prediction. *Journal of Adhesion Science and Technology* 2011; 25: 837–856.

[5] Zhao X, Adams RD, da Silva LFM. Single lap joints with rounded adherend corners: Stress and strain analysis. *Journal of Adhesion Science and Technology* 2011; 25: 819–836.

[6] Adams RD, Atkins RW, Harris JA, Kinloch AJ. Stress analysis and failure properties of carbon-fibre-reinforced-plastic/steel double-lap joints. *Journal of Adhesion* 1986; 20: 29–53.

[7] Stapleton SE, Waas AM, Arnold SM. Functionally graded adhesives for composite joints. *International Journal of Adhesion and Adhesives* 2012; 35: 36–49.

[8] Zhang Y, Sun M, Zhang D. Designing functionally graded materials with superior load-bearing properties. *Acta Biomaterialia* 2012; 8: 1101–1108.

[9] Birman V, Byrd LW. Modeling and analysis of functionally graded materials and structures. *Journal of Applied Mechanics Reviews* 2007; 60: 195–216.

[10] Birman V, Liu YX, Chen CQ, Thomopoulos S, Genin GM. Mechanisms of biomaterial attachment at the interface of tendon to bone. *Journal of Engineering Materials and Technology* 2011; 133: 1–8.

[11] Gay D, Hoa SV, Tsai SW. *Composite Materials: Design and Applications.* CRC Press, Boca Raton, FL; 2003.

[12] Herakovich CT. *Mechanics of Fibrous Composites.* John Wiley & Sons, New York, NY; 1998.

[13] Jones RM. *Mechanics of Composite Materials.* Taylor& Francis, London; 1999.

[14] Tsai SW, Hahn HT. *Introduction to Composite Materials.* Technomic Publishing Company, Lancaster, PA; 1980.

[15] Chai H, Babcock CD. Two-dimensional modeling of compressive failure in delaminated laminates. *Journal of Composite Materials* 1985; 9: 67–98.

[16] Johanneson T, Sjoblam P, Selden R. The detailed structure of delaminated surfaces in graphite/epoxy laminates. *Journal of Material Science* 1984; 19: 1171–1177.

[17] O'Brien TK. Characterization of delamination onset and growth in a composite laminate. *Damages in Composite Laminates* 1982; ASTM STP 775.

[18] Liu S. Delamination and matrix cracking of cross-ply laminates due to spherical indenter. *Composite Structures* 1993; 25: 257–265.

[19] Krüeger R, Minguet PJ. Analysis of composite skin–stiffener debond specimens using a shell/3D modeling technique. *Composite Structures* 2007; 81: 41–59.

[20] Pradhan B, Chakraborty D. Fracture behavior of FRP composite laminates with an embedded elliptical delamination at the interface. *Journal of Reinforced Plastics and Composites* 2000; 19: 1004–1023.

[21] Raju IS. Calculation of strain energy release rates with higher order and singular finite elements. *Engineering Fracture Mechanics* 1987; 28: 252–274.

[22] Rybicki EF, Kanninen MF. A finite element calculation of stress intensity factors by a modified crack closure integral. *Engineering Fracture Mechanics* 1977; 9: 931–938.

[23] Sun CT, Manoharan MG. Strain energy release rates of an interfacial crack between two orthotropic solids. *Journal of Composite Materials* 1989; 23: 460–478.

[24] Li HCH, Dharmawan F, Herszberg I, John S. Fracture behavior of composite maritime T-joints. *Composite Structures* 2006; 75: 339–350.

[25] Di Bella G, Borsellino C, Pollicino E, Ruisi VF. Experimental and numerical study of composite T-joints for marine application. *International Journal of Adhesion and Adhesives* 2010; 30: 347–358.

[26] Chuyang L, Junjiang X. Static pull and push bending properties of RTM-made TWF composite tee-joints. *Chinese Journal of Aeronautics* 2012; 25: 198–207.

[27] Adams RD, Wake WC. *Structural Adhesive Joints in Engineering.* Elsevier Science Publishing Company, London; 1984.

[28] Kinloch AJ. *Adhesion and Adhesives: Science and Technology.* Chapman and Hall, London; 1987.

[29] Tong L, Steven GP. *Analysis and Design of Structural Bonded Joints.* Kluwer Academic Publishers, Boston, MA; 1999.

[30] Matthews FL, Kilty PF, Godwin EW. A review in the strength of joints in fiber reinforced plastics. *Composites* 1982; 13: 29–37.

[31] Marques JB, Barbosa AQ, da Silva CI, Carbas RJC, da Silva LFM. An overview of manufacturing functionally graded adhesives – Challenges and prospects. *The Journal of Adhesion* 2019; 97: 1–35. DOI:10.1080/00218464.2019.1646647.

[32] da Silva LFM, Adams RD. Adhesive joints at high and low temperatures using similar and dissimilar adherends and dual adhesives. *International Journal of Adhesion and Adhesives* 2007; 27: 216–226.

[33] da Silva LFM, Lopes MJCQ. Joint strength optimization by the mixed-adhesive technique. *International Journal of Adhesion and Adhesives* 2009; 29: 509–514.

[34] Pires I, Quintino L, Durodola JF, Beevers A. Performance of bi-adhesive bonded aluminum lap joints. *International Journal of Adhesion and Adhesives* 2003; 23: 215–223.

[35] Fitton MD, Broughton JG. Variable modulus adhesives: An approach to optimized joint performance. *International Journal of Adhesion and Adhesives* 2005; 25: 329–336.

[36] Marques EAS, da Silva LFM. Joint strength optimization of adhesively bonded patches. *The Journal of Adhesion* 2008; 84: 915–934.

[37] Azimi HR, Pearson RA, Hertzberg RW. Fatigue of rubber-modified epoxies: Effect of particle size and volume fraction. *Journal of Materials Science* 1996; 31: 3777–3789. DOI:10.1007/BF00352793.

[38] Wang M, Miao R, He J, Xu X, Liu J, Du H. Silicon carbide whiskers reinforced polymer-based adhesive for joining C/C composites. *Materials and Design* 2016; 99: 293–302.

[39] Barbosa AQ, da Silva LFM, Öchsner A, Abenojar J, Del Real JC. Influence of the size and amount of cork particles on the impact toughness of a structural adhesive. *The Journal of Adhesion* 2012; 88: 452–470.

[40] Barbosa A, da Silva L, Öchsner A. Effect of the amount of cork particles on the strength and glass transition temperature of a structural adhesive. *Proceedings of the Institution of Mechanical Engineers Part L Journal of Materials Design and Applications* 2014; 228: 323–333.

[41] Barbosa A, da Silva LFM, Abenojar J, Del Real J, Paiva RM, Öchsner A. Kinetic analysis and characterization of an epoxy/cork adhesive. *Thermochimica Acta* 2015; 604: 52–60.

[42] Barbosa A, da Silva LFM, Abenojar J, Figueiredo M, Öchsner A. Toughness of a brittle epoxy resin reinforced with micro cork particles: Effect of size, amount and surface treatment. *Composites Part B: Engineering* 2017; 114: 299–310.

[43] Brown EN, White SR, Sottos NR. Retardation and repair of fatigue cracks in a microcapsule toughened epoxy composite—Part II: In situ self-healing. *Composites Science and Technology* 2005; 65: 2474–2480.

[44] Jin H, Miller GM, Sottos NR, White SR. Fracture and fatigue response of a self-healing epoxy adhesive. *Polymer* 2011; 52: 1628–1634.

[45] Brown EN, White SR, Sottos NR. Microcapsule induced toughening in a self-healing polymer composite. *Journal of Materials Science* 2004; 39: 1703–1710.

[46] Carbas RJC, da Silva LFM, Critchlow GW. Adhesively bonded functionally graded joints by induction heating. *International Journal of Adhesion and Adhesives* 2014; 48: 110–118.

[47] da Silva LFM, Öchsner A, Adams RD. Handbook of Adhesion Technology. Springer Science & Business Media, Heidelberg; 2011.

[48] Panigrahi SK, Pradhan B. Adhesion Failure and Delamination Damage Analyses of Bonded Joints in Laminated FRP Composites. PhD Thesis 2007, IIT, Kharagpur.

[49] Carpenter WC. A comparison of numerous lap joint theories for adhesively bonded joints. *Journal of Adhesion* 1991; 35: 55–73.

[50] Zienkiewicz OC. *The Finite Element Method*. Tata McGraw Hill Publishing Company Limited, New Delhi; 2002.

[51] Cheuk PT, Tong L. Failure of adhesive bonded composite lap shear joints with embedded precrack. *Composites Science and Technology* 2002; 62: 1079–1095.

[52] Kairouz KC, Matthews FL. Strength and failure modes of bonded single lap joints between cross-ply adherends. *Composites* 1993; 24: 475–484.

[53] Kayupov M, Dzenis YA. Stress concentrations caused by bond cracks in single-lap adhesive composite joints. *Composite Structures* 2001; 54: 215–220.

[54] Tong L. Strength of adhesively bonded composite single lap joints with embedded cracks. *AIAA Journal* 1998; 36: 448–456.

[55] Kim KS, Yoo JS, Yi YM, Kim CG. Failure mode and strength of uni-directional composite single lap bonded joints with different bonding methods. *Composite Structures* 2006; 72: 477–485.

[56] Qin M, Dzenis Y. Analysis of single-lap adhesive composite joints with delaminated adherends. *Composites Part B: Engineering* 2003; 34: 167–173.

[57] Nimje SV, Panigrahi SK. Interfacial failure analysis of functionally graded adhesively bonded double supported tee joint of laminated FRP composite plates. *International Journal of Adhesion and Adhesives* 2015; 58: 70–79.

[58] Nimje SV, Panigrahi SK. Strain energy release rate based damage analysis of functionally graded adhesively bonded tubular lap joint of laminated FRP composites. *The Journal of Adhesion* 2017; 93(5): 389–411.

[59] Nimje SV, Panigrahi SK. Effects of functionally graded adhesive on failures of socket joint of laminated FRP composite tubes. *International Journal of Damage Mechanics* 2017; 26(8): 1170–1189.

[60] Tsai SW, Wu EM. A general theory of strength for anisotropic materials. *Journal of Composite Materials* 1971; 5: 58–80.

[61] Raghava R, Caddel RM, Yeh G. The macroscopic yield behavior of polymers. *Journal of Materials Science* 1973; 8: 225–232.

[62] Williams JG. On the calculation of energy release rates for cracked laminates. *International Journal of Fracture* 1988; 36: 101–119.

[63] Herszberg I, Li HCH, Dharmawan F, Mouritz AP, Nguyen M, Bayandor J. Damage assessment and monitoring of composite ship joint. *Composite Structures* 2005; 67: 205–216.

[64] Whitcomb JD, Woo K. Analysis of debond growth in tubular joints subjected to tension and flexural loads. *Computes and Structures* 1993; 46: 323–329.

[65] Raju IS, Sistla R, Krishnamurthy T. Fracture mechanics analyses for skin-stiffener debonding. *Engineering Fracture Mechanics* 1996; 54: 371–385.

[66] Dattaguru B, Ramamurthy TS. Venkatesha, Buchholz FG. Finite element estimates of strain energy release rate components at the tip of an interface crack under mode I loading. *Engineering Fracture Mechanics* 1994; 49: 211–228.

[67] Wang JT, Raju IS. Strain energy release rate formulae for skin-stiffener debond modeled with plate elements. *Engineering Fracture Mechanics* 1996; 54: 211–228.

[68] Miravete A, Jimenez MA. Application of the finite element method to prediction of onset of delamination growth. *Journal of Applied Mechanics Review* 2002; 55: 89–106.

[69] Irwin GR. Analysis of stresses and strains near the end of a crack traversing a plate. *Journal of Applied Mechanics* 1957; 24: 361–364.

[70] Raju IS, Crews JH, Aminpour MA. Convergence of strain energy release rate components for edge-delaminated composite laminates. *Engineering Fracture Mechanics* 1988; 30: 383–396.

[71] Tay TE, Shen F, Lee KH, Scaglione A, Di Sciuva M. Mesh design in finite element analysis of post-buckled delamination in composite laminates. *Composite Structures* 1999; 47: 603–611.

[72] Shenoi RA, Violette FLM. A study of structural composite tee joints in small boats. *Composite Materials* 1990; 24: 644–665.

[73] Li W, Blunt L, Stout KJ. Analysis and design of adhesive bonded tee joints. *International Journal of Adhesion and Adhesives* 1997; 17: 303–311.

[74] Li W, Blunt L, Stout KJ. Stiffness analysis of adhesive bonded tee joints. *International Journal of Adhesion and Adhesives* 1999; 19: 315–320.

[75] Panigrahi SK, Zhang YX. Nonlinear finite element analyses of tee joints of laminated composites. *IOP Conference Series: Material Science and Engineering* 2010; 10: 1–7.

[76] Apalak MK. Geometrically non-linear analysis of adhesively bonded corner joints. *Journal of Adhesion Science and Technology* 1999; 13: 1253–1285.

[77] Apalak ZG, Apalak MK, Davies R. Analysis and design of adhesively bonded tee joints with a single support plus angled reinforcement. *Journal of Adhesion Science and Technology* 1996; 10: 681–724.

[78] Apalak MK. Geometrically non-linear analysis of adhesively bonded modified double containment corner joints-II. *Journal of Adhesion Science and Technology* 1998; 12: 135–160.

[79] Apalak MK. Geometrically non-linear analysis of adhesively bonded double containment corner joints. *Journal of Adhesion* 1998; 66: 117–133.

[80] Apalak MK. Geometrically non-linear analysis of adhesively bonded modified double containment corner joint-I. *Journal of Adhesion Science and Technology* 2000; 14: 1159–1178.

[81] Murkherjee A, Varughese B. Design guidelines of ply drop-off in laminated composite structures. *Composites Part B: Engineering* 2001; 32: 153–164.

[82] Kim JS, Kim CG, Hong CS. Practical design of tapered composite structures using the manufacturing cost concept. *Composite Structures* 2001; 51: 285–299.

[83] Lang TP, Mallick PK. The effect of recessing on the stresses in adhesively bonded single lap joints. *International Journal of Adhesion and Adhesives* 1999; 19: 257–271.

[84] Crocombe AD, Adams RD. Influence of the spew fillet and other parameters on the stress distribution in the single lap joint. *Journal of Adhesion* 1981; 13: 141–155.

[85] Harris JA, Adams RD. Strength prediction of bonded single lap joints by non-linear finite element methods. *International Journal of Adhesion and Adhesives* 1984; 4: 65–78.

[86] Sancaktar E, Kumar S. Selective use of rubber toughening to optimize lap-joint strength. *Journal of Adhesion Science and Technology* 2000; 14: 1265–1296.

[87] Patrick RL. *Structural Adhesives with Emphasis on Aerospace Applications*. Marcel Dekker, Inc., New York; 1976.

[88] Raphael C. Variable-adhesive bonded joints. *Applied Polymer Symposium* 1966; 3: 99–108.

[89] Carbas RJC, da Silva LFM, Madureira ML, Critchlow GW. Modelling of functionally graded adhesive joints. *Journal of Adhesion* 2014; 90: 698–716.

[90] Breto R, Chiminelli A, Duvivier E, Lizaranzu M, Jiménez MA. Finite element analysis of functionally graded bond-lines for metal/composite joints. *Journal of Adhesion* 2015; 91(12): 920–936.

[91] Pradhan B, Panda SK. Effect of material anisotropy and curing stresses on interface delamination propagation characteristics in multiply laminated FRP composites. *Journal of Engineering Materials and Technology* 2006; 128: 383–392.

[92] ANSYS, General Purpose Finite Element Software (Version 10.0).

[93] Ravi Chandran KS, Barasoum I. Determination of stress intensity factor solutions for cracks in finite-width functionally graded materials. *International Journal of Fracture* 2003; 121: 183–203.

[94] Erdogan F, Wu BH. The surface crack problem for a plate with functionally graded properties. *Journal of Applied Mechanics* 1997; 64: 449–456.

[95] Shim DJ, Paulino GH, Dodds RH. Effect of material gradation on K-dominance of fracture specimens. *Engineering Fracture Mechanics* 2006; 73: 643–648.

[96] Analas G, Lambros J, Santare MH. Dominance of a asymptotic crack-tip fields in elastic functionally graded materials. *International Journal of Fracture* 2002; 115: 193–204.

[97] Adams RD. Strength predictions for lap joints, especially with composite adherends: A review. *Journal of Adhesion* 1989; 30: 219–242.

[98] Kumar S, Pandey PC. Behavior of bi-adhesive joints. *Journal of Adhesion Science and Technology* 2010; 24: 1251–1281.

[99] Kumar S, Scanlan JP. Stress analysis of shaft-tube bonded joints using a variational method. *Journal of Adhesion* 2010; 86: 369–394.

[100] Temiz S. Application of bi-adhesive in double-strap joints subjected to bending moment. *Journal of Adhesion Science and Technology* 2006; 20: 1547–1560.

[101] Zhou DW, Louca DW, Saunders M. Numerical simulation of sandwich T-joints under dynamic loading. *Composites Part B: Engineering* 2008; 39: 973–985.

[102] Davies GAO, Hitchings D, Ankersen J. Predicting delamination and debonding in modern aerospace composite structures. *Composite Science and Technology* 2006; 66: 846–854.

[103] Dharmawan F, Thomson RS, Li H, Herszberg I, Gellert E. Geometry and damage effects in a composite marine T-joint. *Composite Structures* 2004; 66: 181–187.

[104] Kesavan A, Deivasigamani M, John S, Herszberg I. Damage detection in T-joint composite structures. *Composite Structures* 2006; 75: 313–320.

[105] Baldi A, Airoldi A, Crespi M, Iavarone P, Bettini P. Modelling competitive delamination and debonding phenomena in composite T-joints. *Procedia Engineering* 2011; 10: 3483–3489.

[106] Ozer H, Oz O. Three dimensional finite element analysis of bi-adhesively bonded double lap joint. *International Journal of Adhesion and Adhesives* 2012; 37: 50–55.

[107] Nimje SV, Panigrahi SK. Numerical simulation for stress and failure of functionally graded adhesively bonded tee joint of laminated FRP composite plates. *International Journal of Adhesion and Adhesives* 2014; 48: 139–149.

[108] Panigrahi SK, Pradhan B. Onset and growth of adhesion failure and delamination induced damages in double lap joint of laminated FRP composites. *Composite Structures* 2008; 85: 326–336.

[109] Apalak ZG, Apalak MK, Davies R. Analysis and design of tee joints with double support. *International Journal of Adhesion and Adhesives* 1996; 16: 187–214.

[110] da Silva LFM, Adams RD. The strength of adhesively bonded T-joints. *International Journal of Adhesion and Adhesives* 2002; 22: 311–315.

[111] Dharmawan F, Li HCH, Herszberg I, John S. Applicability of the crack tip element analysis for damage prediction of composite T-joints. *Composite Structures* 2008; 86: 61–68.

[112] Yildirim B, Dag S, Erdogan F. Three dimensional fracture analysis of FGM coatings under thermo-mechanical loading. *International Journal of Fracture* 2005; 132: 369–395.

[113] Jain N, Rousseau CE, Shukla A. Crack-tip stress fields in functionally graded materials with linearly varying properties. *Theoretical and Applied Fracture Mechanics* 2004; 42: 155–170.

[114] Shenoi RA, Hawkins GL. Influence of material and geometry variations on the behavior of bonded tee connections in FRP ships. *Composite* 1992; 23(5): 335–345.

[115] Apalak MK. On the non-linear elastic stresses in an adhesively bonded T-joint with double support. *Journal of Adhesion Science and Technology* 2002; 16(4): 459–491.

[116] Apalak MK, Apalak ZG, Gunes R, Karakas ES. Steady state thermal and geometrical non-linear stress analysis of an adhesively bonded tee joints with double support. *International Journal of Adhesion and Adhesives* 2003; 23: 115–130.

[117] Theotokoglou EE, Moan T. Experimental and numerical study of composite T-joints. *Journal of Composite Materials* 1996; 30: 190–209.

[118] Theotokoglou EE. Strength of composite T-joints under pull-out loads. *Journal of Reinforced Plastics and Composites* 1997; 16: 503–18.

[119] Kumar S. Analysis of tubular adhesive joints with a functionally modulus graded bond line subjected to axial loads. *International Journal of Adhesion and Adhesives* 2009; 29: 785–795.

[120] Kumar S, Scanlan JP. On axisymmetric adhesive joints with graded interface stiffness. *International Journal of Adhesion and Adhesives* 2013; 41: 57–72.

[121] dos Reis MQ, Carbas RJC, Marques EAS, da Silva LFM. Numerical modelling of multi-material graded joints under shear loading. *Journal of Process Mechanical Engineering: Part-E, Institution of Mechanical Engineers* 2020; 234(5): 436–445. DOI: 10.1177/0954408920916112.

[122] Nimje SV, Panigrahi SK. Design and analysis of functionally graded adhesively bonded double supported tee joint of laminated FRP composite plates under varied loading. *Journal of Adhesion Science and Technology* 2015; 29(18): 1951–1970.

[123] Nimje S, Panigrahi SK. Proceedings of 56th Congress of Indian Society of Theoretical and Applied Mechanics. SVNIT Surat, India. 2011. 117–124.

[124] Adams RD, Peppiatt NA. Stress analysis of adhesive bonded tubular lap joints. *Journal of Adhesion* 1977; 9: 1–18.

[125] Thomsen OT. Elasto-static and elasto-plastic stress analysis of adhesive bonded tubular lap joints. *Composite Structures* 1992; 21: 249–259.

[126] Pugno N, Carpinteri A. Tubular adhesive joints under axial load. *Journal of Applied Mechanics* 2003; 70: 832–839.

[127] Esmaeel RA, Taheri F. Stress analysis of tubular adhesive joints with delaminated adherend. *Journal of Adhesion Science and Technology* 2009; 23: 1827–1844.

[128] Alwar RS, Nagaraja YR. Viscoelastic analysis of an adhesive tubular joint. *Journal of Adhesion* 1976; 8: 76–92.

[129] Nemes O, Lachaud F, Mojtabi A. Contribution to the study of cylindrical adhesive joining. *International Journal of Adhesion and Adhesives* 2006; 26: 474–480.

[130] Das RR, Pradhan B. Adhesion failure analyses of bonded tubular single lap joints in laminated fibre reinforced plastic composites. *International Journal of Adhesion and Adhesives* 2010; 30: 425–438.

[131] Cognard JY, Devaux H, Sohier L. Numerical analysis and optimization of cylindrical adhesive joints under tensile loads. *International Journal of Adhesion and Adhesives* 2010; 30: 706–719.

[132] Belingardi G, Goglio L, Tarditi A. Investigating the effect of spew and chamfer size on the stresses in metal/plastics adhesive joints. *International Journal of Adhesion and Adhesives* 2002; 22: 273–282.

[133] Sancaktar E, Nirantar P. Increasing strength of single lap joints of metal adherends by taper minimization. *Journal of Adhesion Science and Technology* 2003; 17: 655–675.

[134] Yan ZM, You M, Yi XS, Zheng KL, Li Z. A numerical study of parallel slot in adherend on the stress distribution in adhesively bonded aluminum single lap joint. *International Journal of Adhesion and Adhesives* 2007; 27: 687–695.

[135] Apalak MK. Elastic stresses in an adhesively bonded functionally graded tubular single-lap joint in tension. *Journal of Adhesion Science and Technology* 2006; 20: 1019–1046.

[136] Apalak MK. Stress analysis of an adhesively bonded functionally graded tubular single lap joint subjected to an internal pressure. *Science and Engineering of Composite Materials* 2006; 13: 183–211.

[137] Ganesh VK, Choo TS. Modulus graded composite adherend for single lap bonded joints. *Journal of Composite Materials* 2002; 36: 1757–1767.

[138] Spaggiari A, Dragoni E. Regularization of torsional stresses in tubular lap bonded joints by means of functionally graded adhesives. *International Journal of Adhesion and Adhesives* 2014; 53: 23–28.

[139] dos Reis M, Carbas R, Marques E, da Silva L. Functionally graded adhesive joints under impact loads. *Proceedings of the Institution of Mechanical Engineers, Part D: Journal of Automobile Engineering* 2021; 235(13): 3270–3281. DOI:10.1177/09544070211004505:1–12.

[140] Zou GP, Taheri F. Stress analysis of adhesively bonded sandwich pipe joints subjected to torsional loading. *International Journal of Solids and Structures* 2006; 43: 5953–5968.

[141] Das RR, Pradhan B. Finite element based design and adhesion failure analysis of bonded tubular socket joints made with laminated FRP Composites. *Journal of Adhesion Science and Technology* 2011; 25: 41–67.

[142] Panda SK, Pradhan B. Analysis of thermoelastic interaction of manufacturing stresses along the interface on interlaminar delamination progression in multi-ply composite laminates. *Journal of Adhesion Science and Technology* 2005; 19: 1305–1323.

[143] Chakraborty D, Pradhan B. Influence of interfacial resin layer on delamination initiation in broken ply composite laminates. *Journal of Adhesion Science and Technology* 2000; 14: 1499–1513.

[144] Glaessgen EH, Riddel WT, Raju IS. Technical report AIAA-98-2022. 1998.

[145] Dattaguru B, Everett RA Jr, Whitcomb JD, Johnson WS. Geometrically non-linear analysis of adhesively bonded joints. *Journal of Engineering Materials and Technology* 1984; 106: 59–65.

[146] Sheppard A, Kelly DW, Tong L. A damage zone model for the failure analysis of adhesively bonded joints. *International Journal of Adhesion and Adhesives* 1998; 18: 385–400.

# Index

Note: **Bold** page numbers refer to tables and *italic* page numbers refer to figures.